GEOWRITING

a guide to writing,

editing, and printing

in earth science

edited by
Robert L. Bates,
Marla D. Adkins-Heljeson,
and Rex C. Buchanan

Fifth Edition

American Geological Institute
Alexandria, Virginia

AGI Staff

Editor: Margaret Oosterman
Book Designer and Page Compositor: Liza Mallard
Proofreader: Edward Wong

Printed on 60 lb. recycled paper
Composed in Palatino using XEROX Ventura Publisher
Cover design by Julie DeAtley
Printed by United Book Press
Printed and bound in the United States of America

Published by the American Geological Institute
Copyright ©1973, 1974, 1981, 1982, 1995
American Geological Institute; all rights reserved

ISBN 0-922152-14-4
Library of Congress Card Catalog Number 95-075923

American Geological Institute
4220 King Street
Alexandria, Virginia 22302-1507
Phone (703) 379-2480   FAX (703) 379-7563

Contents

Acknowledgments

This fifth edition of *Geowriting* is based on earlier editions that were edited by Wendell Cochran, Peter Fenner, and Mary Hill. This edition includes much of their original work; they were not, however, involved in this revision and are not responsible for any errors or changes that may have occurred.

For this fifth edition, the editors wish to acknowledge R. Wayne Davis and Stephanie Webb, who researched and updated entries in the chapter Writer's Guide to Periodicals in Earth Science, and M. Jennifer Sims, who provided updated information for the chapter on Drawings and Photos. Margaret Oosterman provided incisive editing and thoughtful comments, and Julie Jackson devoted considerable time to editing and improving the manuscript.

We appreciate all of these efforts.

Contributions to earlier editions of this book continue to be reflected in the fifth edition. Those contributions include writing, manuscript review, permission to reproduce excerpts, and other help. Those contributors included Gerald M. Friedman, Richard V. Dietrich, Patricia Wood Dickerson, John W. Koenig, Arthur A. Meyerhoff, Thomas F. Rafter, Jr., William D. Rose, Martin Russell, Albert N. Bove, George E. Becraft, A.F. Spilhaus, Jr., Robert McAfee, Jr., Trinda L. Bedrossian, Elizabeth Bennett, Mary Beth Cumming, Jules Braunstein, Kenneth L. Coe, George V. Cohee, Edwin B. Eckel, Josephine F. Fogelberg, William H. Freeman, Douglas M. Kinney, Joel J. Lloyd, Melba W. Murray, Mark Pangborn, Willis A. Shell, Donald G. Turner, Richard J. Vorwerk, Joseph C. Meredith, and Sherman A. Wengerd.

Parts of the chapter Writing for the Media are a slightly different version of material in Geomedia: A Guide for Geoscientists Who Meet the Press by Lisa A. Rossbacher and Rex C. Buchanan (American Geological Institute, 1988). Kenneth K. Landes' paper on abstracts is reproduced by permission from the *Bulletin* of the American Association of Petroleum Geologists, v. 50, September 1966, p 1,992. Excerpts of B. H. Weil's "Standards for writing abstracts," are reprinted by permission from *Journal* of the American Society for Information Science, v. 21, 1970, p. 351; copyright 1970 the American Society for Information Science, Washington, D.C.

Foreword

Peter Fenner, one of the three original editors of *Geowriting*, conceived the idea for this book. The concept for *Geowriting* was his response to reading the report, "Requirements in the field of geology" (1969) by David M. Delo and Robert C. Reeves. That report to the American Geological Institute recommended more stress on facility in writing. *Geowriting* has proved to be as useful to geoscience writers, editors, and students as its creators hoped. New editors in many geoscience organizations routinely receive copies, and *Geowriting* has become the primary resource for earth-science classes in technical writing.

Although the need for facility in writing has not changed in the past 25 years, typesetting, printing, and map-making technologies have changed significantly. The accessibility of personal computers, sophisticated software, desktop publishing, and digital map production are just a few of the technological advances that writers, editors, and students take for granted. The new edition of *Geowriting* takes those changes into account. The section "Writer's Guide to Periodicals in Earth Science" is new to *Geowriting*. Aspiring authors will find information there on more than 150 earth-science journals.

Like the editors of earlier editions, Robert L. Bates, Marla D. Adkins-Heljeson, and Rex C. Buchanan, have been active members in the Association of Earth Science Editors (AESE). In preparing the fifth edition, they have accomplished the challenging task of incorporating new material into a book that evolved from the contributions of many of their peers. Producing a book for other writers and editors is a daunting assignment, and I am certain that the fifth edition will prove to be as useful as its predecessors.

I am grateful to Bob, Marla, and Rex for their good work and to Wendell Cochran, another of *Geowriting's* original editors, who enthusiastically and skillfully guided the book from concept through four editions. We regret that Bob Bates — the 1981 recipient of AESE's Award for Outstanding Editorial or Publishing Contributions — did not live to see the new edition in print.

Julia A. Jackson
Director of Communications
and Publications

Preface

Geowriting is an introduction to writing, editing, and printing in earth science. It will help you cope with a process that begins with writing a manuscript and ends when the work is printed. It is a how-to-do-it manual, an outline, and a guide — a resource intended to give a notion of problems, tools, methods, and references for digging deeper.

Geowriting will be useful to a variety of people:

o Students (undergraduate and graduate) approaching for the first time the possibility of writing for publication.

o Professors and supervisors who foresee that their students or employees will someday write or edit.

o Scientists who publish infrequently and find it difficult to keep in mind all the related mechanical matters.

o Scientists who find themselves — sometimes unexpectedly — appointed or elected to editorships.

The publication process is similar for all types of publications. To simplify explanations in *Geowriting*, we have used writing, editing, and printing an article for journal publication as an example for all publications. The editors' decisions on what to include and emphasize are not meant to be used as standards for all types of publications, or even all journals. They do hope, however, that the examples and discussions will help you make the choices most suitable for your writing.

The book's arrangement generally follows the mechanical order of publication: writing through editing to printing. Writing, editing, and printing are not treated as independent processes, although each can be done without the others. To a craftsman, writing-editing-printing is an interlocking whole.

CHAPTER 1

Before You Write — Preparation

Nearly all scientists enjoy research, the actual doing of science. But many say that the most difficult part of their work is sitting down to write about their research. Some find it hard to get started. They're intimidated by staring at a blank page or an empty computer screen. Others find the entire writing process painful, a necessary evil in the accomplishment of their work. Still, writing is unavoidable if they are to record and communicate their results.

Several techniques will help you get started writing. Once you have started, other methods can help you improve your writing. Quite simply, the only way to get better at writing is to write. Like plumbing or cooking or playing basketball, the more you work at writing, the better you get. It may also help to have someone advise you, telling you which mistakes are easily avoidable and giving you hints to improve your work. But the act of writing will help you clarify your thinking and improve your ability to communicate it.

Choose your writing instrument

Use whatever mechanics appeal most to you. You may want to make the first pass in longhand, or use a typewriter. A personal computer with a word-processing program can make the logistical aspects of writing much easier: you can revise, rearrange, cut, and check spelling and syntax with the touch of a few keys.

Find your writing method

Professional writers use a variety of mechanical techniques and working conditions. Some writers start early in the morning; some work only late at night. Truman Capote claimed that he wrote the first drafts of all his work while lying in bed. Find the method that works best for you and use it to get the product you want.

Make an outline

An outline will help make the rest of the job easier. The level of detail in the outline will depend on your writing experience, the complexity and length of the work, and the publication for which you are writing. Many scientific journals have a standard format that consists of an abstract, a statement of the problem, a review of pertinent literature, a description of research methods and results, and conclusions. It may help you to outline your work with a few words of summary under each of these categories or the categories a particular journal requests.

Paleontologist Stephen Jay Gould is well known for writing about geology, natural history, and the history of science. But even in his technical articles, Gould uses a fairly unconventional writing style, avoiding the usual stylistic conventions and writing in a more narrative style and organization. If you are a good enough writer, you may be able to get away with the same thing. For almost all writers, however, a journal's format may have the advantage of providing an established framework within which you can work.

Think about your audience

A time-honored axiom of writing is to know your audience. You should know the readers' level of understanding of your subject. Some readers will be able to understand complex technical ideas and polysyllabic words without any explanation. After all, the purpose of scientific jargon is to provide a precise and short (though often inelegant) method of communicating technical information. For other audiences, concepts and words may require some definition. You should always keep that level of understanding in mind. Similarly, the audience's level of interest may vary from publication to publication. If you are writing for readers who are unfamiliar with your particular area of expertise, they may require additional explanation to make the relevance of your work clear.

Imagining your audience as one person may help you visualize your readers. That is, think of a person you know who might read your article, such as a colleague or friend. Try to envision this person reading the article, to see if he or she can understand it. Anticipate your reader's questions, both in terms of subject matter and in the technical level of the writing, and then answer the questions. Also, ask yourself what you would like to know as a reader, and then provide that information. Writers should never leave their audience with questions that they could have answered.

You may be your own best audience. You should know, even before you sit down to write, what you want to communicate, which ideas you want to convey. Obviously, you must be interested in and excited about your specific line of research, or you wouldn't have pursued it. You should bring that same interest and excitement to your writing, using your judgment in deciding what is important. If you write about the things that are important to you, you will almost certainly be a better writer.

Put your ideas on paper

Some writers spend hours on the first sentence of a report. They write nothing beyond that first sentence until they are happy with it, no matter how many times they have to rewrite it. For most researchers, however, a better approach is to get something — anything — down on paper without worrying about the grammar, syntax, or misspelled words. The very act of writing will often help you get started. You don't need to worry too much about the way things come out the first time. There is plenty of time to fix them later.

BEFORE YOU WRITE — PREPARATION

- Choose your writing instrument.
- Find your writing method.
- Make an outline.
- Think about your audience.

CHAPTER 2

Preparing Copy

Editors are busy people. Whatever you can do to make their job easier will help get your work into print.

Match your work to a journal

Now that you are ready to write, you will want to find a journal that is interested in publishing your work. Decide which journal is most likely to use it; don't submit your work unless you are familiar with a journal's recent issues. You want to make sure your paper and a journal will make the best fit. For example, the *Journal of Paleontology* and the Geological Society of America's *Bulletin* may both use a paper on Devonian brachiopods, but the *Bulletin* is less likely to use one on a specialized aspect, such as the morphology of productids. The last chapter of this book gives specific information on current periodicals in the earth sciences.

Follow journal or general guidelines for writers

Editorial policies change, so examine recent issues of the journal you have chosen. Do not depend on old issues or on your impression of a journal's content. If a journal has a stylebook or other standard, follow the instructions closely. Some of the rules may seem arbitrary, but follow them anyway — they are designed to fit an editorial system or to meet mechanical requirements imposed by the journal's design or a printer's equipment. Flouting those rules may make extra work for editors. If a journal has no formal suggestions to authors, apply common sense and use a stylebook, such as the *Suggestions to Authors of the Reports of the United States Geological Survey* or *The Chicago Manual of Style*.

Hard copies. Stylebooks may specify that manuscripts be submitted on good quality white bond paper (some may even specify the weight). Editors must work with the paper, and they object to paper too stiff to handle, too thin for convenience, too shiny for easy reading, or too hard or slick to take sharp, clear pencil marks. Papers with easy-erase surfaces are particularly troublesome.

Computer printouts should be easy to read and photocopy. Some printers produce hard-to-read type: for example, a lowercase *g* may look like a 9. Some space words so that editors cannot easily count words or characters for estimating printed pages. Many editors actually ban certain types of printouts, notably dot-matrix. Laser printouts are more than adequate. Check a publisher's stylebook for specifications, or send a sample page.

Electronic copies. Many journals request that manuscripts be submitted in electronic form as well as on paper. Confirm with a journal the computer system and software that was used so that you can submit a readable diskette or one that can be translated. Submitting a manuscript via electronic mail may also be acceptable and convenient. A paper copy should also be sent to verify the electronic transmission.

Typing. All copy should be typed, double-spaced, with standard indentions, at least four-centimeter wide margins on all sides, and approximately equal line lengths. That rule applies to all copy, no exceptions. It includes titles, bylines, author identification and affiliation, abstracts, quoted matter, footnotes, tables, lists of references — everything. Double-space everything. Editors need the space between the lines and in the margins for editorial marks. Also, even spacing and margins help editors estimate how much space your work will occupy in the publication. Try not to use a proportional-space typewriter or printer, which makes copy-fitting — estimating space — all but impossible.

Do not add extra spaces between paragraphs or sections. Triple spacing is acceptable, but be sure to use the same spacing throughout. Single spacing is never acceptable. Avoid devices such as single-spacing the abstract, long quotations, or references in an attempt to simulate smaller type. Do not break a word at the end of a line, because an editor or typesetter may mistake the hyphen for a part of the spelling.

Use one side of the paper only. At the top of each page, type a short identification tag in case pages get misplaced — one word, such as your last name, and the page number: *Jones 1, Jones 2, Jones 3.*

Estimating space. Some editors welcome an outline of a prospective manuscript because both a writer and editor can then discuss manuscript length in terms of the number of words, typescript pages, and printed pages. As a guide, one 8 1/2-inch by 11-inch, two-column printed page set in 10-point (elite on a typewriter) type equals about three and one-half double-spaced typewritten pages, excluding figures.

Names. Provide complete names for people. A text reference to a person's name should match a bibliographic reference. Do not abbreviate names or terms, especially journal names, unless a journal specifically requires it. That job is for the editor, who should not have to find out whether your *Geol.* means *Geologic, Geological,* or *Geologists.*

Numbers. If possible, avoid built-up fractions: $\dfrac{a-b}{3(a+c)bx}$

They are hard to set in type and waste space. Use case fractions: *(a-b)/3(a+c)bx*. If you have typed equations using a word-processing program, find out how to save the file so that the equation format can be retained when the file is accessed. Make sure you send hard copy so that equations can be rebuilt if necessary.

Most science journals use the international metric system. Check to see if the journal you chose does.

Tables. Avoid rules, which are horizontal or vertical lines. Remember that the editor can add needed rules quickly and neatly; taking unneeded ones out is much more difficult and time-consuming. Use enough space between columns to make the meaning clear but not so much as to make lines of numbers hard to follow with the eye.

Footnotes. Avoid using footnotes if at all possible. Many journals will not allow footnotes and will incorporate the material into the text.

Verification. Double-check all spellings, quotations, references, equations, formulas, and arithmetic. It is an editor's duty to check some of these, too, but an editor is under time constraints and usually working on several manuscripts at once. The author is responsible for providing correct information and mathematics.

Permissions. If you plan to quote extensively from published material or to reproduce another author's photographs, tables, or diagrams, you must get permission in writing from the copyright owner. Many publishers have standard permission forms, which simplify and speed this task. For more information, please refer to chapter 11, Editing and Proofreading.

Avoid trade names, such as Plexiglas, Geodimeter, and Xerox. Carelessness in using them can result in sharp letters and worse from attorneys. Dictionaries usually indicate trade names, which should be capitalized. The trademark or registration symbol is not necessary.

Artwork. Photographs and other artwork must be identified in such a way that they will not be damaged. You may write the information on a separate piece of paper and tape it to the back of the artwork. Be sure to include location, scale, and any pertinent credit. If you have artwork done on a computer, check with the editor to see if the journal has the same graphics applications. If not, find out how to save your graphics and text files so that they can be accessed. For more information on artwork, please see chapter 7, Artwork.

Design. Do not attempt to specify design or to mark typographic style except for foreign words or phrases, such as names of species. Design and style are an editor's job.

Number of copies. For most editors, the original and one copy will suffice. Some editors require multiple copies, however, so check. Note that some photocopying machines use paper that is hard to write on; test all paper with a soft black pencil. Be sure to keep a copy for yourself. Let a journal know if you want your work back and enclose an envelope with your name, address, and sufficient postage.

PREPARING YOUR MANUSCRIPT

- Select an appropriate journal.
- Match your work to a journal.
- Follow journal or general guidelines for writers.

CHAPTER 3

Getting It Written

Scientific research is not complete until the results have been published. Therefore, a scientific paper is an essential part of the research process. Therefore, the writing of an accurate, understandable paper is just as important as the research itself. Therefore, the words in the paper should be weighed as carefully as the reagents in the laboratory. Therefore, the scientist must know how to use words. Therefore, the education of a scientist is not complete until the ability to publish has been established. (From Robert A. Day, *How to Write and Publish a Scientific Paper*)

We assume that you have decided where to submit your paper and have the appropriate style manual or sheet of instructions. We also assume that you have written an outline from which to proceed.

Decide on a title

It is wise to devote some care to this label for your product, because readers deserve an accurate statement of an article's contents. Two requirements are involved:

o The title should tell what the paper is about.

o The title should not be long and cumbersome.

Meeting these requirements will also help make your paper easy to cite by future workers. Remember that increasing use is being made of computer-oriented indexing and searching techniques. When indexed, most of the words in your title should help a reader search the literature by key words. Editors sometimes must modify titles of papers; you can help both editor and reader by keeping the title of your article brief and specific. Avoid such words as *introduction, principles, selected, investigations,* and *recent.*

Express the title clearly

After deciding on the title content, be sure to express the title clearly. A succession of words that seems to make perfect sense to you may not be clear to others. In *Abandoned Copper Mine Subsidence Study*, the first word, an adjective, can modify any of the following four words, all of which are nouns. Presumably, the author did not want to refer to an abandoned study. The title may be improved by using a modifying phrase: *Subsidence Study of an Abandoned Copper Mine*. The real subject of the paper, however, is not the study but the subsidence. Why not call it *Subsidence at an Abandoned Copper Mine?*

Another title starts with the words *Submersible Observations.* To the author, both these words are nouns, but to many readers *submersible* is an adjective meaning capable of functioning under water. Relating *submersible* to *observations* is difficult. *Observations from a Submersible* would have done the job nicely.

Follow writing guidelines

The following section provides some general guidelines for clear writing. The explanations are not intended to be comprehensive but rather to emphasize specific points. The chapter Reference Shelf lists many excellent references that give detailed information.

Declarative sentences. A straightforward, declarative sentence is the most useful vehicle in scientific writing. A subject (person, process, or thing) acts on or affects an object or result. Such a sentence is a normal forward-action unit, in which the verb is in the active voice:

> The rocks / contain / plagioclase.
>
> Diagenetic changes / may destroy / the open porous structure.

A number of variations on this basic framework exist. In both sentences, the verbs (*contain, may destroy*) may have adverbs as modifiers, for example, *commonly contain,* and *may ultimately destroy.* In the second sentence, the subject and object are modified by adjectives (*diagenetic, open, porous*). A phrase adds further meaning: *the open porous structure of the diatomite.*

Nonrestrictive clauses. A nonrestrictive, or *which* clause, which is not essential to the meaning of the sentence but adds to it, may be included. Such clauses are set off by commas:

> *Many of the rocks contain plagioclase, which has normal zonation.*

Restrictive clauses. You may need a restrictive *that* clause, which is essential to the sentence meaning and requires no commas:

> *the open porous structure that typifies normal diatomite*

Passive voice. An alternative structure turns the sentence around:

> Plagioclase / is contained / in the rocks.
>
> The open porous structure/may be destroyed/by diagenetic changes.

The verbs are in the passive voice. That is, the subject of the sentence is being acted upon and is thus passive. Most good writers view the passive voice unfavorably; the active voice is inherently more dynamic and usually shorter. If the rocks contain plagioclase, that's the way to say it. But we don't lay down an absolute antipassive rule. If the sentence subject is the texture and structure of diatomite, it is logical to place the subject first and use the passive voice. You would then discuss the diagenetic changes.

Beware of the passive voice with weak verbs, such as *is seen, is found, was made,* and *was done.* Avoid passive verbs when preparing an abstract (discussed in a later section).

Subject-verb agreement. The verb in a sentence must agree with its subject in number, even though subject and verb may be separated:

> *A collection of museum-grade minerals, rocks, and fossils was available.*

By putting the verb close to the subject, you will be more likely to notice any disagreement between subject and verb.

There is/There are. Starting sentences with the words *There is* or *There are* is permissible, but in general this usage should be avoided. *There is an abundance of fossils* means *Fossils are abundant.*

Sentence length. Try to vary the length and complexity of your sentences. No reader likes a paper full of short, choppy sentences, or long sentences with numerous subordinate clauses and other decorations. To test how well you are doing, when you finish a paragraph or a page read it out loud. You should be aware of the sounds that "words make on paper," as E.B. White put it. Rework your sentences until they sound right.

Paragraph length. Just as there is no set length for a sentence, there is no predetermined length for paragraphs. When writing for scientific journals, authors should write paragraphs that focus on one idea. They should use as many sentences as necessary, then begin a new paragraph when shifting focus or ideas. Thus, lengthy paragraphs (about eight, 10, or 12 sentences) are permissible in scientific writing, even though they may be frowned on in other types of less formal writing.

Transitions. As you shift focus from one paragraph to the next, be sure to include transitions: words or phrases that connect paragraphs. Such transitions make life easier for your readers by telling them how you are changing directions, how the discussion to come is related to the subject they were just reading. Sometimes a single word provides that transition. Words such as *however* or *although* at the beginning of a sentence wave a flag to the reader that you are about to qualify or perhaps even contradict what was written in the preceding paragraph. Sometimes phrases or even entire sentences are necessary to perform a transition. For example, the first sentence in the third paragraph of this chapter acts as a transition sentence:

> *After deciding on the title content, be sure to express the title clearly.*

The first part of the sentence describes the purpose of the preceding paragraph, and the second part clues the reader to the rest of the paragraph. That transition sentence ties the two paragraphs together.

Avoid fancy writing

This sentence is from a letter sent out by a geological consulting firm:

> *To properly categorize and document current investigative methodologies, an intensive data gathering effort must be initiated.*

How does that sound when you read it out loud? Is that the way we normally talk? Of course not. The sentence has dressed up a simple thought to look impressive. *To properly categorize and document* (we will forgive the split infinitive) apparently means to determine or find out. *Current investigative methodologies* is an elaborate way of saying methods or techniques now being used. The sentence winds up with a typical passive-voice construction:

> An intensive data gathering effort must be initiated.

By whom? In this final flourish, we may omit *intensive*, as presumably no one would make a lackadaisical data-gathering effort. Note the hyphen, not in the original, but needed because *data-gathering* is a unit modifier of *effort*. Translated into English, the sentence becomes

> *To find out what methods are now being used, we need to obtain information.*

or even

> *We need more data on methods now in use.*

These two suggested revisions are shorter than the original and may be said aloud without embarrassment.

A type of fancy writing that should be avoided is the use of long terms based on classical Latin instead of shorter equivalents from the Anglo-Saxon. *We initiated measurement of the adjacent arenaceous strata* means *We started measuring the nearby sandstone beds*. There are exceptions, when *approximately* sounds better than *about*, or *subaqueous* better than *under water*. One author, however, wrote that dune sands were moved about by *aeolian mechanisms*. He meant wind.

Don't let your modifiers dangle

Every adjective, adverb, phrase, or clause modifies some term in a sentence. An obvious rule is that the modifier must go near, preferably next to, the term modified. This sentence was in a book:

> *Beginning 4 billion years ago, the authors show how microbes invented all of life's essential systems.*

To avoid implying excessive age for the authors, the sentence might well have been rephrased:

> *The authors show how microbes, beginning 4 billion years ago, invented all of life's essential systems.*

Keep modifiers next to what they modify, lest absurdity result. A special case is the adverb that floats unattached:

> *Hopefully, the job will be done this week.*

There is nothing in this sentence for *hopefully* to modify. It should be replaced by *We hope* or *It is hoped*. Floating adverbs do not belong in serious writing. The author who wrote *This study was gratefully supported by the National Research Council* should have said *I am grateful to the National Research Council for its support*. Of course, such adverbs can be used correctly. We speak hopefully when we say you won't misuse them.

Nouniness and how to avoid it

No doubt you will agree that *field* is a fine upstanding noun. So is *oil*. Put them together and you have *oil field*, two nouns end to end. No problem here. We can even take a third noun, *giant*, and place it in line, making *giant oil field*. This phrase hardly poses any difficulty, but from here on things get progressively messier. We have the production record of the field, which we designate *giant oil field production record*. This phrase contains some interesting data, which we analyze, giving us a *giant oil field production record data analysis*. We then construct a diagram based on these data — a *giant oil field production record data analysis diagram* — and naturally conclude with a preliminary interpretation. So the paper is entitled *Giant Oil Field Production Record Data Analysis Diagram Preliminary Interpretation*.

Readers should never be asked to fight their way through such clotted prose — nine nouns and an adjective, all in a heap. The cure for such writing is the phrase. By using a few prepositions, we can recast the title into English that is immediately understandable:

> *Preliminary Interpretation of a Data-Analysis Diagram of the Production Record of a Giant Oil Field*

or, if you prefer:

Data-analysis Diagram of the Production Record of a Giant Oil Field: Preliminary Interpretation.

We hyphenate *data-analysis* to make it a unit that modifies *diagram*, and we convert the other terms into prepositional phrases. Nothing can be done to make that title a model of graceful prose, but at least we can make it comprehensible.

Each of the following titles has appeared in geological literature. Can you translate them into English?

Canadian Superior Harmattan Area Gas Processing Plant Sulphur Recovery Exemption Application

Multiple Pulse Incoherent Scatter Correlation Function Measurements

Heavy Mineral Magnetic Fraction Stream Sediment Geochemical Exploration Program

Unit modifiers

We have already mentioned a type of three-word expression in which the first two words modify the third — as in *three-word expression*. Although putting a hyphen between the first two words clearly aids the reader, this bit of help is often omitted. For example, a tectonic lineament roughly coextensive with a part of the 38th parallel has been referred to as the *38th parallel lineament*. This phrase is poor usage; it implies that there are a lot of parallel lineaments and this is the 38th. The expression is given its correct meaning by a hyphen: *38th-parallel lineament*. The first two words make a unit that modifies the third.

Such expressions as the following require a hyphen: *high-level terrace, rare-earth element, low-angle fault, mean-dip map.*

Sometimes editors will remove hyphens for reasons that are not obvious to authors. If you think a hyphen is needed, discuss its value with the editor. Its inclusion may be needed.

Sexist language

Sentences that refer only to one sex when they could equally apply to both can often be corrected (and shortened) by using plural constructions. For example:

The geologist should use the reflection and refraction profiles when he is uncertain of the dip of underlying formations.

might be rewritten:

Geologists should use reflection and refraction profiles when they are uncertain of the dip of underlying formations.

Instead of writing

When the geologist begins, he or she should visit the site immediately,

write

Geologists should begin by visiting the site immediately.

Spelling

If you have trouble remembering that *consistent* is spelled with an *e* and *resistant* with an *a*, don't feel bad; our language is full of teasers like that. Help is always available. The most obvious source is the dictionary. Keep one nearby and don't be embarrassed to use it. The *Glossary of Geology* and the *Dictionary of Geology* are also helpful. If you use a word processor, use the spelling checker with it.

For quick reference, you may want to make a list of words that give you trouble. The following list seems to bother some geologists. .

symmetrical (two *m*'s)

consistent, persistent

desiccate (one *s*, two *c*'s)

discernible

eustasy, isostasy (no *c*'s)

fluorite, fluorspar

liquid, liquefy

occurred, occurrence (two *c*'s, two *r*'s)

permeable, permeability

phosphorus

predominant, resistant

soluble

Mohs

Punctuation

The best way to learn about this subject is to note how various punctuation marks are used in material that you read, and to use references such as those listed in the chapter Reference Shelf. Counsel here is brief.

Comma. A comma, which seems to give the most trouble, marks a slight pause in the flow of words. For example, in the preceding sentence, the *which* clause stands out from the main sentence and is enclosed by commas.

Semicolon. A semicolon marks a slightly longer pause than a comma.

Period. For a full stop, or period, William Zinsser remarks that there isn't much to say about it except that "most writers don't reach it soon enough." (*On Writing Well*)

Colon. The most common use of a colon is to tell what's coming.

Dashes. A long dash separates:

> *No vestige of a beginning — no prospect of an end.*

Some writers also use dashes to denote a more emphatic form of parentheses:

> *The sea left behind layers of shale and limestone — generally shale in the shallow sea and limestone where it was deeper — along with deposits of coal.*

A hyphen connects, as in unit modifiers such as *low-angle fault*, and between syllables at the end of a line of type.

Apostrophe. An apostrophe denotes possession. The general rule is an apostrophe goes inside an *s* if the possessor is singular, outside if plural: *the rock's age, the pebbles' average size.* Depending on your stylebook, you may write *the 1980's* or *the 1980s*.

Quoting

You may sometimes want to quote the words of another writer. You should repeat these words verbatim, enclose them in quotation marks, and cite the source from your list of references, for example *(Snarf, 1984)*. Copyright permission is necessary for quotations of several paragraphs or a page or more. Or you may rephrase another author's remarks in your own words — as long as you give the source.

References

You should have a list of references at the end of your article. The entries must be in the format required by the journal to which you send your paper. Journal editors are fanatics in this matter, so be sure to follow the instructions in the journal's guidelines to authors. If guidelines are unavailable, use a recent journal issue as a model. If you aren't writing for a publication but are preparing a document such as a company report, adopt a format for the references and use it consistently. Spell authors' names correctly and verify all information. Listing references is loaded with opportunities for error.

Software packages are now available that let you compile references as you work, then make style adjustments according to your avenue of publication. These packages can help you make sure that your reference style is appropriate and consistent.

TO GET YOUR MANUSCRIPT WRITTEN

- Decide on a title.
- Express the title clearly.
- Follow writing guidelines.
- Avoid fancy writing.

CHAPTER 4

Revise!

One widespread misconception about professional writers is that they get it right the first time. Not many do. The difference between ordinary writing and good writing often lies in revision. Revision is far more radical than checking copy for grammatical mistakes, punctuation problems, and syntax errors. It requires looking at every word, every sentence, and every paragraph to make sure that each is where it is supposed to be and says what you want it to say. With every reading, ask these questions:

o Does the sequence of sentences make sense in a paragraph, or should some sentences be moved?

o Does the order of the paragraphs flow logically from one to the next?

o Are the paragraphs connected by transitions?

Revise as much as necessary. Revision often requires reworking a manuscript several times before it is finished. As you revise, consider the following guidelines.

Use action verbs

Verbs are the action part of a sentence; good writing usually uses active, visual verbs, not nondescript verbs, such as *is, was,* or *have.* One mode of revision is to underline all the verbs. Underlining may help you pay special attention to them and show if you are using vivid active verbs or dull passive ones.

Also, look to the readability of your sentences. If a verb is placed far from a subject, your readers may have trouble following the sentence meaning. Placing the verb close to the subject may help.

Vary sentence length

Don't use all long or all short sentences. A short sentence on the heels of a long one often has substantial impact.

Shorten manuscript

Look for ways to shorten your manuscript. Research has shown that short words are usually easier to understand than long ones; short sentences are easier than long ones; and short paragraphs are easier than long ones. Shorter versions also get your ideas across in less space, making readers more likely to read them.

Seek colleague advice

You may want to show your manuscript to colleagues before you submit it, partly for technical reasons but also for readability. The sentences that seem the clearest to you may cause your readers confusion. The results may be, at least at first, somewhat painful. Writing is such a personal process — our words become so much a reflection of ourselves that they are like flesh and blood — that criticism can be hard to take. Most writers learn, however, that such criticism is most profitable and heads off mistakes later. You should understand that dealing with critical analysis is part of the professionalization of your work; that is, criticizing your writing can lead to improvement, the same way that you may benefit from critical suggestions about research techniques or other facets of your work. If your colleagues can help you in the revision process — shortening, polishing, improving — take all the help you can get.

REVISING

- Use action verbs.
- Vary sentence length.
- Shorten manuscript.
- Seek colleague advice.

CHAPTER 5

On Citing Sources

When writing a technical paper, you must identify information you have used from other sources and the sources themselves. This accountability gives professional credit where credit is due.

Avoid trouble for yourself, editors, reviewers, and publishers: Find out if a particular citation style is required. Most journals do require a specific style, and those styles vary widely. So use the journal stylebook if one exists. If one is not available, study examples in a recent journal issue.

If all else fails, devise a style suitable to the subject matter and use it consistently. Aim for simplicity and utility. Long bibliographies are a good place to see examples of citation styles.

Rules for reference citations may seem complex, but they are necessary. Much editorial time is spent cleaning up reference lists, showing how poorly the rules are understood — and how much editors value them.

The following checklist shows the elements that should be included in a list of references:

o Author's surname and given name or initials. List the author's name the way the author uses it; or the editor's surname and given name.

o Publication date. The year usually suffices.

o Title of work, such as a journal article or book chapter.

o Special category of publication, such as abstract, edited work, editorial, or photograph.

o Title of publication, such as book, journal, symposium issue, or field guide.

o Volume number. Include part number and issue number, if the pages are not numbered consecutively, such as in a publication in which each article, chapter, or section is paginated separately.

o Name of publisher.

o Place of publication. Needed for nonserial publications.

o Page reference. Use the total number of pages if the entire publication was used, or list the pages of the part that was used in your work.

o Information that will enable a person to locate the reference if it is unpublished.

It is a good idea to spell out rather than abbreviate titles, particularly to avoid confusion when citing foreign and uncommon sources of literature. Space saved by abbreviations, many editors contend, is not worth the frustration for readers who cannot decipher abbreviations or for editors who must check and correct them.

Verify all information. Listing references is loaded with opportunities for error.

References should be listed at the end of your paper and should include only those sources you have cited in the body of the paper. Thus, a list of references differs considerably from a bibliography, which should include references pertinent to your topic, even if they are not cited in your paper.

It has been estimated that 15,000 serial publications include papers pertinent to geology. GeoRef, the database produced by the American Geological Institute and used to compile the *Bibliography & Index of Geology*, includes about 4,000 serial titles annually; some 2,000 of these include about 95 percent of the world's geological literature. GeoRef is an excellent source to search for references when preparing a paper. The database is available in many academic libraries, as a set of CD-ROMs, and on-line. GeoRef is easily searchable and often is the first step in any literature search.

Other databases commonly used for literature searches in the earth sciences are the Science Citation Index and the National Technical Information Services listings compiled by the Department of Commerce and covering reports from federal agencies and contractors.

CHAPTER 6

Abstracting the Essence

The most-read part of a paper may be its abstract. Effective abstracts are concise, summarize conclusions and recommendations, and are amenable to computer storage and retrieval. In terms of number of readers, an abstract is easily the most essential part of a technical paper.

To help explain what an abstract is, we have included two views of abstracts. View one, by Kenneth K. Landes, was published in the *Bulletin* of the American Association of Petroleum Geologists in 1966 (Vol. 50, No. 9, p. 1992). View two is an excerpt from "Standards for writing abstracts" by B.H. Weil.

View one

A Scrutiny of the Abstract, II

ABSTRACT

A partial biography of the writer is given. The inadequate abstract is discussed. What should be covered by an abstract is considered. The importance of the abstract is described. Dictionary definitions of "abstract" are quoted. At the conclusion a revised abstract is presented.

For many years I have been annoyed by the inadequate abstract. This became acute while I was serving a term as editor of the *Bulletin* of the American Association of Petroleum Geologists. In addition to returning manuscripts to authors for rewriting of abstracts, I also took 30 minutes in which to lower my ire by writing "A Scrutiny of the Abstract." This little squib has had a fantastic distribution. If only one of my scientific outpourings would do as well! Now the editorial board of the Association has requested a revision. This is it.

The inadequate abstract is illustrated at the top of the page. The passive voice is positively screaming at the reader! It is an outline, with each item in the outline expanded into a sentence. The reader is told what the paper is about, but not what it contributes. Such abstracts are merely overgrown titles. They are produced by writers who are either (1) beginners, (2) lazy, or (3) have not written the paper yet.

To many writers the preparation of an abstract is an unwanted chore required at the last minute by an editor or insisted upon even before the paper has been written by a deadline-bedeviled program chairman. However, in terms of market reached, the abstract is the *most important part of the paper*. For every individual who reads or listens to your entire paper, from 10 to 500 will read the abstract.

If you are presenting a paper before a learned society, the abstract alone may appear in a preconvention issue of the society journal as well as in the convention program; it may also be run by trade journals. The abstract which accompanies a published paper will most certainly reappear in abstract journals in various languages, and perhaps in company internal circulars as well. It is much better to please than to antagonize this great audience. Papers written for oral presentation should be *completed prior to the deadline for the abstract*, so that the abstract can be prepared from the written paper and not from raw ideas gestating in the writer's mind.

My dictionary describes an abstract as "a summary of a statement, document, speech, etc...." and that which *concentrates in itself the essential information* of a paper or article. . . . May all writers learn the art (it is not easy) of preparing an abstract containing the *essential information* in their compositions. With this goal in mind, I append an abstract that should be an improvement over the one appearing at the beginning of this discussion.

ABSTRACT

The abstract is of utmost importance, for it is read by 10 to 500 times more people than hear or read the entire article. It should not be a mere recital of the subjects covered. Expressions such as "is discussed and "is described" should *never* be included! The abstract should be a condensation and concentration of the *essential information* in the paper.

View two

An abstract, as defined here, is an abbreviated, accurate representation of a document. The following recommendations are made for the guidance of authors and editors, so that abstracts in primary documents may be both helpful to their readers and reproducible with little or no change in secondary publications and services.

Make the abstract as informative as the document will permit, so that readers may decide whether they need to read the entire document. State the purpose, methods, results, and conclusions presented in the document, either in that order or with initial emphasis on findings.

For various reasons, it is desirable that the author write an abstract that the secondary services can reproduce with little or no change. These reasons include the economic pressures on the secondary services caused by continuing increases in the volume of scholarly publication; the need for greater promptness on the part of the secondary services in publishing information about the primary literature; and the growing value of good authors' abstracts in computerized full-text searching for alerting and information retrieval.

In the proposed standard the term *abstract* signifies an abbreviated accurate representation of a document without added interpretation or criticism and without distinction as to who wrote the abstract. Thus, an abstract differs from a brief *review* of a document in that, while a review often takes on much of the character of an informative or informative-indicative abstract, its writer is expected to include suitable criticism and interpretation. While the word *synopsis* was formerly used to denote a résumé prepared by the author, as distinct from an abstract (condensation) prepared by some other person, this distinction no longer has real meaning.

Types of abstracts

An abstract should be *informative*; that is, it should present quantitative and qualitative information. Space limitations may influence the amount of information you can present but not the quality. Informative abstracts are especially desirable for texts describing experimental work and documents devoted to a single theme. Discursive or lengthy texts, however, such as broad overviews, review .papers, and entire monographs, may permit an abstract that is only an *indicative* or descriptive guide to the type and contents of a document. A combined *informative-indicative* abstract must often be prepared when limitations on the length of the abstract or the type and style of the document make it necessary to confine informative statements to the primary elements of the document and to relegate other aspects to indicative statements.

Abstracts should not be confused with the related, but distinct, terms *annotation, extract*, and *summary*. An *annotation* is a note added to a title or other bibliographic information of a document to comment or explain, such as the notes on the references shown in the chapter entitled Reference Shelf in this book. An *extract* signifies one or more portions of a document selected to represent the whole. A *summary* is a restatement within a document (usually at the end) of its salient findings and conclusions and is intended to complete the orientation of a reader who has studied the preceding text. Because other vital portions of the document (for example, the purpose and methods) are not usually condensed into a summary, the term should not be used synonymously with *abstract*.

Format

For long documents, such as reports and theses, an abstract generally should not exceed 500 words and preferably should appear on a single page. Most papers and portions of monographs require fewer than 250 words. Fewer than 100 words should suffice for notes and short communications. Editorials and Letters to the Editor often will permit only a single-sentence abstract.

Begin an abstract with a topic sentence that is a central statement of a document's major thesis, but avoid repeating the words of a document's title if the title is nearby.

In abstracts specifically written or modified for secondary use, state the type of the document early in the abstract if the document type is not evident from the title or publisher or if it will not be clear from the remainder of the abstract. Explain either the author's treatment of the subject or the nature of the document, for example, theoretical treatment, case history, state-of-the-art report, historical review, report of original research, or literature survey.

Write a short abstract as a single, unified paragraph; use more than one paragraph for long abstracts, for example, those in reports and theses. Write complete sentences, using transitional words and phrases for coherence.

Avoid terms, acronyms, abbreviations, and symbols that may be unfamiliar to your readers unless you define them the first time they occur in the abstract. Include short tables, equations, structural formulas, and diagrams only when necessary for brevity and clarity. Try not to cite references.

A well-prepared abstract enables readers to identify the basic context of a document quickly and accurately, to determine its relevance to their interests, and thus to decide whether they need to read the entire document. Readers for whom the document is of fringe interest often obtain enough information from the abstract to make their reading of the whole document unnecessary. Therefore, every primary document should include a good abstract. Secondary publications and services that provide bibliographic citations of pertinent documents should also include abstracts if at all possible.

CHAPTER 7

Drawings and Photos

Drawings, photographs, and maps are the usual artwork for articles. *Line drawings* are created by solid black lines, so they can be printed with text material. *Photographs* (both black-and-white and color) are continuous-tone images in which the images have a full range of tones from black to white. For printing, a screening process reduces these tones to evenly spaced dots of varying shape, size, and number. The density of the dot areas varies in direct proportion to the intensity of the image area they represent. *Maps* can be either line drawings or color images.

Preparation of geologic maps is discussed in chapter 8, Readying the Map. Most geologic maps are printed in four-color process from separate plates for each color, prepared from drawings or negatives that a cartographer makes. However, it is possible to scan hand-colored maps electronically and to make color separations from the scan for process color printing. Full-color maps can also be produced by computer digitization and printing through electrostatic means (where light is reflected onto an electrically charged drum and toner is fused to the paper in the areas retaining a charge, such as a laser printer) or by pen or drum plotters. Making photographic copies of hand-colored maps is possible but expensive.

Good graphics are a must. Take the same care with artwork that you do with an original piece of research. Good graphics and good research make possible a magnificent presentation.

Photos

Almost everyone can take photographs. For that reason, perhaps, a photograph is the form of illustration most abused. If you want to support your paper with good photos, take care that you are proficient with the camera or hire a qualified photographer.

Taking photos. Each photograph should be accurate and clear. Most non-professional photographers stay too far away from their subjects. If you know a geologic unit thoroughly, communicate this intimacy in your photograph. Get close to the subject, or plan to crop closely when you make the print.

Wait for good natural light if you are taking exterior photographs, or add light if you cannot wait. A poorly lighted photograph seldom shows the subject well and will be a poor illustration when printed.

If you do not know the general requirements for good photo composition, consult a how-to art or photography book. With care, you can take photographs that will improve your scientific contribution. If you do not feel that you can handle the photography needed for your manuscript, hire a technical or scientific photographer. The gain in artistic quality for your manuscript will probably be well worth their charge.

To find out what size photographic print to submit, ask the editor of the publication you are writing for or check the publication stylebook. Generally, editors prefer photographs that are slightly larger than the printed version will be. Allow an editor to determine the percentage of photo reduction or enlargement. Prints should generally have a high gloss to display a maximum amount of detail. Photographs also can be electronically scanned and turned into digital data; currently, quality of digitized photographs is often comparable to that of the original. Check with your editor on which is preferred. Advances in technology will make possible a digitized file of photographs that at least equal the quality in an original, and a digitized file has the capability of being enhanced electronically.

Handling photos. Cleanliness, neatness, and care are watchwords with all artwork. Once you have a good photographic print suitable for reproduction, treat it with great care. Don't dent it with paper clips, and never staple artwork. Identify it so that an editor can tell which photograph matches which caption. Refrain from writing on the back in such a way that the writing comes through as embossed lines on the front. To be safe, use rubber cement or tape to attach a caption (with your name, date, illustration number, and article title) on the back of each photo. Be sure the top of the photograph is marked clearly.

Do not write on the face of artwork, even if you would like lettering to appear on the printed picture; when the photograph is screened for printing, your lettering will be screened, too. If you want to draw lines or identify objects in a photograph, do the drawing and lettering on a transparent plastic overlay. The overlay needs to be clearly marked to assure proper registration — that is, so that marks on the overlay will be printed exactly where you want them to be printed. Generally, little circles with crosses in them, called bull's-eyes, are used for registration. Art stores stock them in adhesive rolls and in sheets. Place register marks outside the margins of both pieces of art, one aligned exactly over the other. Three marks on each sheet should guarantee registration.

Art stores also stock letters and numerals in a variety of styles and sizes, as well as symbols and patterns, suitable for doing professional lettering on your overlay or drawing if you are expected to submit final copy. Once you have made an overlay or drawing, treat it with care. If a drawing has been made with anything that can smear, such as pencil or charcoal, spray it with a fixative. Cover all artwork with a protective sheet of paper. Remember that any writing, even erased pencil lines drawn for lettering, or writing on a sheet of paper on top of a photograph, may leave pressure marks that will show when printed.

Cropping

Be sure to obtain a clear photograph in good focus, cropped (trimmed) as you would like it. If the photofinisher has not cropped it enough, you can indicate further cropping to an editor by making a paper mask to go around the photo on all sides. The mask will not interfere with the system used to mark cropping limits and reductions for the printer. Suggestions for cropping follow:

o Enhance a subject. Crop a photo to stress its purpose. If a caption reads "The rock is horizontally layered," then the best cropping would probably be horizontal.

o Crop tightly to a subject, maintaining good composition and photographic interest.

o Crop to improve composition. Reduce distractive elements, focus on a subject, and improve balance and relationship of design elements in a photo.

o Crop to remove photographic blemishes.

o Watch the scale. A graphic scale is by far the best type of scale. If an object indicates a scale in a photograph, don't crop out that object unless you show the scale in some other manner. If you must use a caption that gives a mathematical scale for the photo, be sure that the scale is maintained in printing. You will need to recalculate a scale if a different size photo is used.

o Do not crop people out of a photo if they contribute to reader interest in the subject or if they indicate scale or depth of field. (Do not, however, allow people to distract from the purpose of a photo.)

o Vary the shapes of photos to add interest to a page. Unless uniformity is a deliberate means of design, avoid it. Uniformity does not ensure quality.

o In general, make sure that scenics, or photographs encompassing large areas, will be printed larger than close-ups.

Scales

A scale in a photograph should be identified in a caption (for example, 20-meter tree). If a photograph has no scale, you may add a *graphic scale* on an overlay or state the *mathematical scale* (amount of magnification or reduction) in a caption. Your editor should take mathematical scales into account when calculating printing enlargements and reductions. But editors sometimes forget or miscalculate — so graphic scales are greatly preferred.

Line maps and figures, illustrations that do not need to be converted into dots for printing, should be drawn clearly to present a neat, legible appearance when printed. In general, they should be drawn larger than they will be when printed — 15 percent larger works best. That way, lines will appear cleaner but the size of the drawing is not changed dramatically. Any lines or letters must remain bold and clear after being reduced.

If you can afford it, you may want to have your drawing reduced photographically or by photocopier to show you exactly how it will look when printed. Some editors prefer to use a reduced photographic negative and a positive print rather than original artwork. If your artwork is publishable without changes, this is an excellent option; if not, you may be asked for originals. If your budget cannot bear the cost of a negative, you can have your drawing reduced by a photocopier to check your work. A printed illustration will be much better than a photocopy, but don't let faith in the printing process blind you to faults in your drawing.

To save wear and tear on original artwork, you may wish to submit only photocopies of your drawings (but never of the photographs) until your article has been accepted for publication. Photocopies, however, are *not* suitable for publication. If you have created graphics on a computer, find out if the journal uses the same graphics software that you use. If so, you can send a copy of your graphic on a diskette. An editor can then revise a graphic if necessary and place the figure in electronic form in a page layout. Always include a printed copy of a graphic in the event that electronic transfer fails. If you have hand-drawn graphics, an editor may still have them digitized through scanning so that they can be placed electronically.

Size and shape

The proportions your drawing or photograph will have if reduced can be determined in several ways. The easiest way is to use a reducing lens or a simple proportional-scale device (both are available at art stores). Other ways use graphic and mathematical methods. Such methods will tell you if your reduced drawing or photograph is the right shape, but not if it will be legible.

Although as author you should plan your drawings to fit the format of the journal selected, you should not mark reductions on illustrations. An editor is responsible for marking the illustration size reduction for a printer.

Some journals require a border around each illustration. Check to see if a border is required; if not, let the editor decide whether your artwork will be improved by a border. (Too much bordering can kill an illustration.)

Figuring reductions

Figuring reductions for artwork is generally done by an editor or designer. Very rarely, an author will have to figure reductions. If you must calculate reductions, the graphic method or a mathematical method should help you:

Graphic method. The graphic method uses a diagonal line and a ruler to figure reductions. If you assemble your work on a light table, you will be able to see easily the different layers of pages you are using.

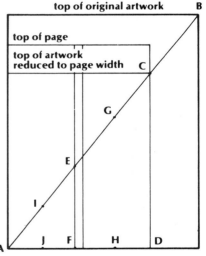

o Make a dummy of the page shape used in the journal, as shown in the figure. Note that the figure shows a page with a two-column layout.

o Place a tissue overlay over the dummy page.

o If an illustration is to be within page or column margins, place the lower left corner of the illustration at A, under the dummy. On the tissue overlay, very lightly draw a diagonal line from A to B, at the upper right corner of the illustration.

o To reduce an illustration to a page width, measure the illustration height from C (where the margin intersects line to the bottom of the page at D. That measurement is the height of the reduced illustration.

o To reduce an illustration to a column width, measure the illustration from E to F.

o Any intermediate width can also be measured (for example, at G-H or I-J).

Mathematical method. Set up a simple ratio-and-proportion equation, for example:

present width : present height   reduced width : reduced height

5 inches : 4 inches   2.5 inches : y inches

Solve the equation: $5y = 4(2.5)$. The reduced height (y) will be 2 inches.

Writing captions

Check to see if a journal requires captions for all artwork or uses a specific caption style. A single word or phrase may be sufficient for some journals; others may require complete sentences. Writing captions has no hard-and-fast rules. Maintain consistency in style, whether you use single words or simple declarative sentences. Perhaps you need a caption that states *Figure 1*; perhaps it should state *Figure 1: Map shows the southeast corner of the northwest part of the Eldorado 15-minute quadrangle, Boulder County, Colorado*. In identifying features on a photograph, use either letters or numbers on an overlay and explain them in a caption. Give all necessary credits. You do not need to identify people in the distance in photographs unless they provide necessary information. Courtesy requires that you identify people in a foreground. You must supply a typed list of your captions with your manuscript. Key each caption to the correct piece of artwork so that the editor does not have to worry about mixing them up.

It is both an editor's and an author's job to make sure captions are as clear, complete, and concise as possible. The author should mark the position of each illustration in the margin of a manuscript, so that pertinent text will be printed as closely as possible to the illustration. Before you send the text and artwork to press, check again to be sure that each piece of artwork and each overlay is identified and clearly marked for reduction. You must be especially careful to mark each illustration so that the printer cannot possibly confuse them.

Label each illustration outside the margin, where the label will not be printed. As author, you must identify yourself, your article, and your illustration; for example, *Robert F. Smith, Ross ice sheet, figure 1, green overlay*. An editor can add the journal name and other necessary information.

Printing in full color

Full-color printing is expensive, because color is complicated and difficult to reproduce accurately. A good example is minerals, which are often colorful to the eye, but their colors are difficult to reproduce. A photographic reproduction of a mineral rarely does justice to the mineral; consequently, a printed illustration is rarely an accurate representation of the original mineral.

The main reason that printed color does not always show accurately colors of actual objects or colors seen in color transparencies (or color slides) and color prints lies in the different ways color is seen, photographed, and printed. When colors you see are photographed, the colors are reproduced chemically. When these chemically produced colors are translated into printer's ink, a different system of color is used. A printing process called full-color process printing reproduces the full range of colors. Full-color printing uses four colors, called process colors. When printed, the process colors appear as tiny dots of solid color, which are combined in various sizes and patterns to duplicate the full range of colors in an original print or transparency. Colors are mixed optically, by the eye of the viewer.

Two types of color photographs can be made: photos made from color negatives, from which you can have photographic prints made in color or in black and white; and color transparencies, which are see-through positives. Either can be used to make printing plates after the colors are separated photographically or electronically. If you are using color photographs, ask the editor which kind of color photo will best fit the journal's production methods: a color print or a color transparency. If you plan to publish in black and white, take black-and-white photographs. A color photograph can be printed in black-and-white, but the printed image will not be as clear as one made from a black-and-white photograph.

Hand-drawn color artwork for full-color process printing may be sent to the printer separated by the artist, in which case you will submit overlays for the various colors and shades of color to be printed, or unseparated. Colors in an unseparated illustration (for example, a painting or a color drawing) must be separated to be printed.

A color scanner may be available to digitize your artwork. Check with your editor to find out if you can use color in a publication and in what form the editor wants your work.

Preparing slides for lectures

If you are like most scientists, you will not make many color photographs or slides with publication in mind. Instead, you will make them to illustrate talks and papers to be given to students, colleagues, or the general public. You will, no doubt, want to add other slides to them to complete your lecture.

In preparing slides, keep in mind the proportion of a 3-inch by 5-inch card, vertical or horizontal. If you are using 8 1/2-inch by 11-inch paper to prepare your illustration to be shot as a slide, use 18- to 24-point type (0.25-0.33 inch), preferably Helvetica or another clear, easy-to-read, sans-serif typeface. Use uppercase and lowercase letters, because all uppercase letters are extremely difficult to read in a slide. Line weights should be proportionally heavy.

When you prepare illustrations, keep in mind that they may be slides. Rarely will a printed diagram, map, or drawing become a good slide without extra work. References listed in the chapter Reference Shelf provide guidance for creating effective slides. The following suggestions by Duncan Heron for preparing and presenting a slide talk are reprinted from the *AGI Data Sheets* (American Geological Institute, third edition, 1989).

The purpose of a slide talk is to communicate one or more ideas to an audience. Data are presented as maps, graphs, charts, and photographs. Gathering scientific data is often a long and expensive process. Prepare your presentation as carefully as you gathered your data, and consider the following aspects: copy, production, and showmanship.

Copy

Design and art production are separate but closely related parts of preparing slide copy. Try to follow these rules for a design:

- Keep the design simple with only one idea per slide.
- Plan your design at a 2:3 ratio, using either 6" x 9" or 8" x 12" paper.
- Keep the format horizontal.

Once you have a design, show it to a colleague and briefly explain the point of the diagram. Then remove the diagram and ask questions. You should quickly know if your design works.

You or an artist may prepare the art, using many methods, including a computer (CAD); dot-matrix graphs do not, however, make good slides. Slide copy must be bold: Letter size, line weights, and symbols should be large. Remember these guidelines for art production:

- Letter size is a function of copy size and viewing distance. Using Pratt and Ropes' (1978, *35-mm Slide*, AAPG) assumption that maximum seating distance is six projector screen widths, to determine minimum letter size, take the longest dimension of the slide image area and multiply by 2. This gives the type point size. Convert to inches by multiplying point size by .014. Some examples follow.

Longest image dimension	Point Size	Inch Size
9	18	0.25
12	24	0.33
18	36	0.50

- Line weights should be thicker than those in journal illustrations. A minimum of 1/32" is used with a 6" x 9" image area. Make prime data lines two times heavier, or 1/16".
- Bullet size for 6" x 9" should be no smaller than 0.10". Use a larger size for the most important data points.
- Letter style should be simple and uniform. Avoid script and gothic, outlined, and similar fancy letter styles.
- Color should be used to emphasize the important point of the slide and to replace cross-hatching and other patterns. Colors used on maps or sections should conform to U.S. Geological Survey usage for rock types.

Production

Turning the finished artwork into a slide is essentially a copying process. Try to follow these rules for manual production:

- Use a copying stand equipped with 3200 K lights.
- Use a 35-mm SLR camera in the manual mode.
- Use daylight slide film with an 80A blue filter, or use tungsten film and no filter.
- Align the copy so that it fills the frame and is square with the border of the frame.
- Use anti-glare glass if reflections are a problem.
- In the manual mode, determine the exposure on a gray card (a neutral test card with 18% reflectance).
- Make extra exposures by bracketing one stop above and one stop below the gray card reading.

Electronic slide production also is widespread. Presentation software, film recorders, slide scanners, and software to enhance photographs are all available and constantly improving.

Showmanship

A polished slide talk is a result of well-planned slides, careful integration of the slides with oral presentation, and practice. One cannot over-emphasize practice. Some guidelines follow:

- Clean and preload your slides.

- Make certain the slide on the screen corresponds to what you are saying. The slide acts as a prompt-card. When you change to another point, change to a related slide. If you don't have a related slide, leave the screen black or use a neutral gray blank slide.

- Never return to a previous slide. Use a duplicate.

- Look at the audience, not the screen.

- Do not overuse a pointer.

- Avoid the phrase "I apologize for this slide, but..." If you must apologize for a slide, do not use it. If you must use a poor slide, do not apologize.

PREPARING YOUR ILLUSTRATIONS

- Use clear photos in good focus.

- Handle photos with care.

- Label each illustration outside the margin.

- Include a graphic or mathematical scale in captions.

- Write clear, complete, concise captions.

- Use uppercase and lowercase letters in slides.

CHAPTER 8

Readying the Map

Had we but world enough and time, we could produce maps slowly and beautifully as we once did. But the world is shrinking and time speeding, and today's demands are for maps today. The usefulness of maps is the major reason for increasing demand. Geologic maps are sophisticated tools, capable of presenting millions of bits of data in an extremely efficient manner.

Before you undertake to make a map, consult with an editor to find out a journal's size and printing limitations; an editor should check with a printer. At most state geological surveys, geologists may consult with the cartographic supervisor as soon as they receive a map-making assignment, and long before starting field work. The supervisor supplies the best possible base map to use. Most likely, the base map will be especially compiled for the job and printed in green on heavy translucent Mylar film. The map can be drawn directly on this base, without fear that it will stretch or shrink out of scale. After compilation, several techniques can be used to produce a finished map. Three methods are described.

Preprinting preparation

To prepare a map for conventional full-color process printing, begin by making an Ozalid print (a blueprint) of the map for checking and editing. When the map is ready for a cartographer, the green base can be photographically eliminated, leaving only an author's black lines representing the geological data to be transferred. If needed, the green lines can be retained as check points; because they are green, they are easily separable from a geologist's black lines. Decisions are made on colors, screens, and patterns so the cartographer can proceed. It is necessary to correctly register the different layers of the map (such as the base, culture, geology, etc.) at this time for precise color registration during printing. Registry holes can be punched at a printer's or cartographer's shop. Bull's-eyes are applied to all sides in an irregular fashion to show clearly the top and bottom of a map. The map is transferred to a special film that is coated with an opaque layer, called a scribe coat, then photographically transferred to a film with a coating that can be peeled off to expose areas. The cartographer scribes — or cuts the film away — to make negatives of a map, one sheet of film for each color and pattern. These negatives must be registered, or the lines of the map will not match up and will be distorted in printing.

Camera separation

In camera separation of hand-colored maps for four-color process printing, the initial procedure is similar to that for the conventional method: Authors compile field data on scale-stable green- or blueline copies of base maps. After the map has been approved, however, the author covers the linework or data with an overlay of frosted film and adds color by pencil, acetate film, or other mediums. When the overlay (which becomes the camera-ready copy) has been colored, checked for accuracy, and reviewed and approved by the author, four negatives are prepared. One is a combination negative of the base map (probably screened), lettering, and other positive work. The others are negatives prepared by photographing the colored overlay three times, once each with a filter of the primary colors for photography or light (blue, green, and red). Each overlay is made into a plate, which is printed in black, yellow, magenta, and cyan. Again, several elements make up the complete map: the base (usually area or county borders, screened so as to not detract from the black-line data), the frosted-film overlays for black-line data and lettering, and one overlay for each color. Each overlay must be registered with the others.

The following steps must be taken to prepare registered artwork:

1. Register a frosted-film overlay by hand-punching registry holes and adding bull's-eyes to a scale-stable base map. A graphic arts department of a photo supply store should have register pins and punch. The overlay will become the black-line base art or database. The database will usually have been drafted in ink, but it can also be scribed (where film is cut away to create a negative). Inked lettering and type can be put directly on the data overlay unless they are on paper or other opaque material. If so, another overlay must be prepared.

2. A separate overlay for type must be exactly registered to the database overlay. The type itself can be a mixture of strip film, cut-up typewritten paper, or photocopy held in place with transparent tape. (For uniform results, you should stick to a single medium.)

3. The frosted-film overlay for color is prepared last; then, after registration, hand coloring can begin.

Computerized cartography

Personal computers make drafting maps much quicker than hand-drawn methods. Computer-generated maps are made with commercially purchased mapping applications (using existing data sets) or a scanner and a drafting program.

To generate maps using computers, map data are entered into a computer in numerical form, or an existing map is digitized — converted into numbers. Data sets generated on a digitizing table from a paper or Mylar map are entered into a computer. Numbers are manipulated and combined with additional base-map files, such as county boundaries and township-and-range lines. Once data are digitized, they can be displayed using map-making software packages that are commercially available. A map can be edited on a computer screen, or it can be plotted, using such computer-driven plotters as drum, pen, or electrostatic plotters, and edited on paper. Corrections are entered on a computerized version.

Once a computerized map is completed, a final product can be plotted and printed. Depending on the type of plotter, the map may be produced in black-and-white, in a limited number of colors, or in full-color. These maps can then be printed. However, problems may arise during printing, because patterns applied by a plotter may conflict with screen angles applied by a printer. Computers can also be used to speed printing. For example, some types of plotters can be used to produce scribe coats.

Generating maps by computer allows much easier updating or correction, so that new versions of a map can be produced quickly. In addition, by using Geographic Information System (GIS) software, maps can be generated at a variety of scales or with any combination of thematic elements.

Page-size, black-and-white maps for book and journal publication can be scanned into digital form to be used as a template in a drafting program. A template is a file that acts as a starting point for other illustrations. After a map is scanned, a file can be opened in a drawing program and drafted. Patterns and screen tints, explanation text, and other labels can be added to areas on a map. A map can be saved in a format that can be imported into a page composition program. All corrections can be made electronically and a floppy disk used for printing. Small, colored flat-color (spot color) maps can also be produced this way, with color separation for four-color process printing done electronically at a printer.

Limits of size and paper

Before settling irrevocably on a certain style or size of a map, you should be aware of certain limits. Paper size and press size need to be considered. Large maps can obviously present much more detailed information than smaller maps. Large maps, however, besides being unwieldy, do not endure; they are often folded incorrectly so many times that soon they disintegrate into a mass of crinkles. A major publisher of geologic maps has a press that will accommodate paper up to 42 inches by 58 inches. That size map is so large that one must crawl across it to see detail at the center; it may serve in a library or classroom, but it is almost unmanageable in the field.

With the advent of computers, the same map can be produced at a variety of scales. The scale of a map should reflect the scale at which data were collected and meant to be displayed. For example, if data for a geologic map were originally compiled at a scale of 1:100,000, the same map can be generalized and reproduced at 1:500,000 with a loss in detail but no loss in accuracy. The same 1:100,000 map cannot be reproduced at the more detailed 1:24,000 scale without giving a false sense of the map's accuracy.

Although authors seldom have control over the kind of paper used, editors do — they should insist on long-lasting, acid-free paper in durable, acid-free pockets. Paper containing much sulfur disintegrates quickly on library shelves.

Symbols

Commonly used symbols for topographic maps are included in U.S. Geological Survey publications. There is no universally recognized standard for geologic map colors or symbols, such as contacts, lithologic types, and structural relationships. To choose the most appropriate symbols for your map, consider the following suggestions:

o Review the U.S. Geological Survey system for geologic map symbols, which is discussed in *Suggestions to Authors* (See chapter 8, Reference Shelf, "Stylebooks" section). Map symbols are also listed in chapter 8, "Symbols" section.

o Review other maps drawn to similar scales — if any exist — to show features similar to yours that have been mapped in other locations.

o Most important, include a complete legend or explanation that defines all the symbols you use in your map, including the significance of various tints, lettering styles, and other overprints used.

Checklist

Maps, legends or explanations, and cross sections should be checked for the following:

o Completeness.

All units are labeled.

Formations in the legend and cross section are also on the map.

Geographic locations mentioned in the text are shown on the map.

o Correctness.

Plotting is correct.

Names are spelled correctly.

Spelling on the map agrees with that in the text.

Locations described in the text and shown on cross sections agree with the map.

Numbers are correct.

Names and ages in the legend agree with those on the map and in the text.

Symbols in the legend are also on the map and those on the map are also in the legend.

Colors and patterns in the map and cross sections match those in the legend.

Dips drawn in a cross section agree with those on the map.

Scale is appropriate to data presented.

Colors or black-and-white are used correctly to show data.

o Title. A concise, yet complete, title should include subject and location. Include the state, possibly the county, and perhaps the country, for example, *Geologic map of Kern County, California, U.S.A.*, or *Map of the tectonic features of the United States*.

o Author, compiler, contributors, and sources of data or base maps.

o Scale, preferably both graphic and numerical.

o Contour interval and datum, where appropriate.

o Fieldwork or compilation dates.

o Publication or copyright date.

o Publisher name and location.

o Distributor name and location, if needed.

o Identification of series and sheet if more than one sheet is used, for example, *Economic Mineral Investigation 3: sheet 2*.

o Arrows showing true and magnetic north, and declination.

o Sponsors, drafters, and cartographers (so labeled). These contributors need to be labled as such (for example, Dow Jones, cartographer) so that they are clearly differentiated from the map's authors.

o Geographic reference points and grids, such as township-range grid; longitude-latitude; Army Map Service grid.

o Lines of cross sections.

o Identification in upper right corner, for example, *Map Sheet 54, Dubuque quadrangle*.

o Color or pattern block and explanation of all units on the map; most maps will include age designations and abbreviated lithology.

o Explanation of symbols.

o Horizontal scale, if different from the map or separated from it.

o Vertical scale or vertical exaggeration.

o Orientation.

For United States national map-accuracy standards, write to the U.S. Geological Survey, Reston, Va. 22092. A published map meeting these requirements should say in the legend, *This map complies with national map-accuracy standards*. Given the litigious nature of today's society, a disclaimer may also be appropriate for certain maps, particularly those that may be a source of information for economic or health-related decisions. Disclaimers can clarify the standards of accuracy used in a map's compilation, and they can state that the publisher is not responsible for decisions based on a map.

PREPARING YOUR MAPS

- Check journal size and printing limitations.
- Choose appropriate map symbols.
- Include a complete legend.
- Use checklist when verifying maps, legends or explanations, and cross sections.

CHAPTER 9

Rules for Geologic Names

Things, time, places, and events form the basic framework for writing geological reports. Over the years, stratigraphers, structural geologists, and other specialists have named these categories to aid communications. Certain conventions have been adopted so that names have a similar meaning for everyone.

Stratigraphic classification

The following information about names and their usage has been adapted from several sources, primarily the U.S. Geological Survey (USGS), the American Commission on Stratigraphic Nomenclature, and the International Subcommission on Stratigraphic Classification.

Stratigraphic classification is a systematic organization of rock strata into units with reference to any or all of the characteristics or properties they possess, especially lithologic character, fossil content, age and time relationships, seismic and magnetic properties, electric-log characteristics, mineral assemblages, lithogenesis, and environments of deposition or formation. Lithozones, biozones, chronozones, mineral zones, and other zones usually designate minor stratigraphic intervals in their classification categories. A zone name is capitalized when used as a formally named unit.

In general, formally named stratigraphic units are capitalized when used in their entirety. For example, in the Hutchinson Salt Member of the Wellington Formation, all proper names are uppercase; if only Hutchinson salt is used, the "s" in salt is lowercase. The Names Committee of the USGS prefers that plural formal names retain capitalization, contrary to most style manual rules for plurals of formal nouns; for example, when referring to both the Lansing Group and the Kansas City Group together, preferred USGS usage is Lansing and Kansas City Groups rather than Lansing and Kansas City groups.

o Lithostratigraphic units are bodies of rock strata characterized by a unique lithology or a combination of lithologic types not present in adjacent units. The formation is the fundamental mappable unit of lithostratigraphic classification. Formally named formations are capitalized, for example:

> *the Dundee Limestone, a Middle Devonian formation in the Michigan basin*

the Flathead Sandstone, a sandstone deposited on a Cambrian sea floor in what is now Wyoming

A formation may be subdivided into members, such as the formally named *Ferron Sandstone Member in the Mancos Shale*, or the informally named *green shale member*. Members may be divided into beds, for example, such as the Fence-post limestone bed in the Pfeifer Shale Member. Two or more associated formations having significant lithologic features in common may be included in a group, such as the *Cisco Group*.

o Biostratigraphic units are designated by fossil content or by paleontologic character that differentiates them from adjacent units. The basic unit is the biozone. Formally named biozones are capitalized:

the Heterostegina Assemblage Zone of the Gulf Coast area

the Cardioceras cordatum Range Zone

o Chronostratigraphic units are bodies of rock formed during some specified interval of geologic time. Examples of formally named units are *Phanerozoic Eonothem, Paleozoic Erathem, Silurian System, Middle Silurian Series*, and *Tonawandan Stage*. Corresponding geochronologic units, representing the time during which the chronostratigrahic units were formed, are *eon, era, period, epoch*, and *age*. Thus, rocks of the Silurian System were deposited in the Silurian Period of the Paleozoic Era.

The stratigraphic column in this chapter is organized to allow generalized correlation between measured time and occurrence of time-rock units.

Guides to usage

A series of guides to U.S. stratigraphic usage have been prepared by the American Commission on Stratigraphic Nomenclature. The latest revision, entitled *Code of stratigraphic nomenclature*, is available from the American Association of Petroleum Geologists. The guide is an explicit statement of principles and practices for classifying and naming stratigraphic units. Some conventions from the guide follow.

o Words used in formal names of rock-stratigraphic (lithostratigraphic) units are capitalized:

Ash Creek Group, Chinle Formation, Kirtland Shale, Church Rock Member, Sonsela Sandstone Bed

Informal names, as an unnamed sandstone bed in the Chinle Formation, are not capitalized: *a Chinle sandstone bed*.

o Capitalization of formal and informal names of time (geochronologic) and rock-time (chronostratigraphic units) follows similar conventions:

Paleozoic Era, Devonian Period, Cenomanian Stage, Cenomanian Age

but

Devonian time, Devonian age, and Paleozoic age

The last example is a mixture of formal and informal time terms.

o Formal names of zones are capitalized, except for the italicized or underscored species name of a plant or animal:

Bulimina excavata Concurrent-range Zone

o The terms *lower, middle,* and *upper* are capitalized when they describe formal series subdivisions of a system; *early, middle,* and *late* are the corresponding formal (and therefore capitalized) time terms:

Upper Cretaceous rocks were deposited in Late Cretaceous time.

o The terms *lower, middle,* and *upper* are lowercased when they describe informal chronostratigraphic units. The corresponding informal time terms *early, middle,* and *late* are lowercased when they describe subdivisions of eras; formal series of the Tertiary, such as *lower Pliocene* or *early Pliocene*; and provincial series, such as *lower Atokan* or *early Atokan*. Distinction between formal and informal chronostratigraphic terms are found in *Suggestions to Authors.*

Proposed lithostratigraphic units should be described and defined clearly for easy recognition. An intent to introduce a new name and the important factors that led to discrimination of the unit should be clearly stated. A definition should give the geographic or other feature from which the name is taken and the specific location of one or more representative sections near the geographic feature. Specific reference to location in section, township, and range, or other land divisions should be included. Thickness, lithology, color, and age of the unit should be given.

U.S. Geological Survey bulletins

U.S. Geological Survey Bulletins 896, 1056-A, 1056-B, 1200, 1350, 1502-A, 1520, 1535, 1564, 1565 (all lexicons and reference works on stratigraphic nomenclature) provide definitions and published references to formally named geologic units in the United States. These publications are also available on CD-ROM. For questions on specific place names, check with the Board on Geographic Names (c/o U.S. Geological Survey, Reston, VA 22092). Fossil names should adhere to conventions of the International Rules of Zoological Nomenclature and the International Code of Botanical Nomenclature.

This chart shows general correlations between absolute time—in numbers—and the occurrence of formally named rock-time units. Dates at system or series boundaries are in millions of years (approximations, of course). The smaller divisions are stage names except for the Cambrian series listed. European (left) and North American classifications given. Brackets mark major orogenic climaxes.

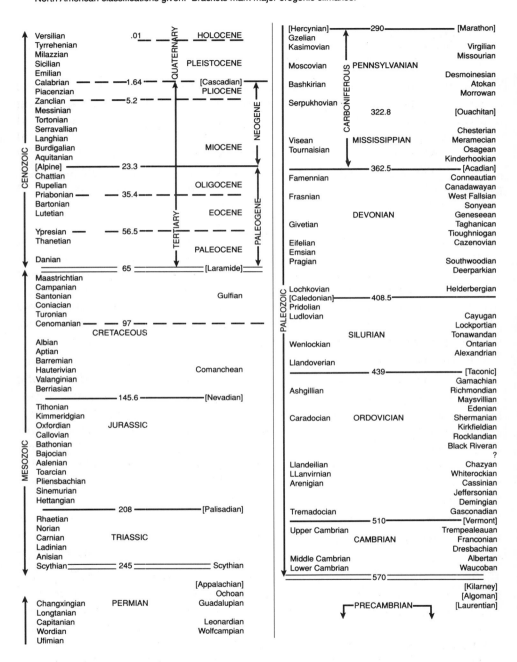

CENOZOIC

European	Date	System/Series	North American
Versilian	.01	QUATERNARY — HOLOCENE	
Tyrrhenian			
Milazzian		PLEISTOCENE	
Sicilian			
Emilian			
Calabrian	—1.64—	PLIOCENE	[Cascadian]
Piacenzian			
Zanclian	—5.2—		
Messinian			
Tortonian		NEOGENE	
Serravallian			
Langhian			
Burdigalian		MIOCENE	
Aquitanian			
[Alpine]	—23.3—		
Chattian		OLIGOCENE	
Rupelian			
Priabonian	—35.4—		
Bartonian		EOCENE	PALEOGENE
Lutetian		TERTIARY	
Ypresian	—56.5—		
Thanetian		PALEOCENE	
Danian			
	—65—		[Laramide]

MESOZOIC

European	Date	System/Series	North American
Maastrichtian			
Campanian			
Santonian			Gulfian
Coniacian			
Turonian			
Cenomanian	—97—	CRETACEOUS	
Albian			
Aptian			
Barremian			
Hauterivian			Comanchean
Valanginian			
Berriasian			
	—145.6—		[Nevadian]
Tithonian			
Kimmeridgian			
Oxfordian		JURASSIC	
Callovian			
Bathonian			
Bajocian			
Aalenian			
Toarcian			
Pliensbachian			
Sinemurian			
Hettangian			
	—208—		[Palisadian]
Rhaetian			
Norian			
Carnian		TRIASSIC	
Ladinian			
Anisian			
Scythian	—245—		Scythian

European	Date	System/Series	North American
			[Appalachian]
			Ochoan
Changxingian		PERMIAN	Guadalupian
Longtanian			
Capitanian			Leonardian
Wordian			Wolfcampian
Ufimian			

PALEOZOIC

European	Date	System/Series	North American
[Hercynian]	—290—		[Marathon]
Gzelian			
Kasimovian			Virgilian
			Missourian
Moscovian		PENNSYLVANIAN	
			Desmoinesian
Bashkirian			Atokan
			Morrowan
Serpukhovian			
	322.8	CARBONIFEROUS	[Ouachitan]
			Chesterian
Visean		MISSISSIPPIAN	Meramecian
Tournaisian			Osagean
			Kinderhookian
	—362.5—		[Acadian]
Famennian			Conneautian
			Canadawayan
Frasnian			West Fallsian
			Sonyean
			Geneseean
Givetian		DEVONIAN	Taghanican
			Tioughniogan
Eifelian			Cazenovian
Emsian			
Pragian			Southwoodian
			Deerparkian
Lochkovian			Helderbergian
[Caledonian]	—408.5—		
Pridolian			
Ludlovian			Cayugan
			Lockportian
			Tonawandan
Wenlockian		SILURIAN	Ontarian
			Alexandrian
Llandoverian			
	—439—		[Taconic]
			Gamachian
Ashgillian			Richmondian
			Maysvillian
			Edenian
Caradocian		ORDOVICIAN	Shermanian
			Kirkfieldian
			Rocklandian
			Black Riveran
			?
Llandeilian			Chazyan
LLanvirnian			Whiterockian
Arenigian			Cassinian
			Jeffersonian
			Demingian
Tremadocian			Gasconadian
	—510—		[Vermont]
Upper Cambrian			Trempealeauan
		CAMBRIAN	Franconian
			Dresbachian
Middle Cambrian			Albertan
Lower Cambrian			Waucoban
	—570—		
			[Kilarney]
			[Algoman]
		PRECAMBRIAN	[Laurentian]

Information compiled from U.S. Geological Survey *Suggestions to Authors* (7th ed., rev. and ed. by Wallace R. Hansen, 1991; *Geological Time Table* (3rd ed., comp. by F. W. B. Van Eysinga, rep. 1983, Elsevier Scientific publishing Co.); and *A Geologic Time Scale 1989* (by W. Brian Harland, Richard L. Armstrong, Allen V. Cox, Lorraine E. Craig, Alan G. Smith, and David G. Smith, 1990, Cambridge University Press).

CHAPTER 10

Judgment by Peers

Every manuscript submitted for publication should represent an author's best efforts. Ideally, it has been written, rewritten, set aside for a cooling period, rewritten again, and polished. Even if all these steps are taken and a manuscript is an author's best effort, it will still benefit from technical review. An author is too close to the work; a fresh, objective look by someone else is essential to spot errors in fact or reasoning, methodological problems, inconsistencies, or poor presentation that obscures what an author has tried to convey.

Technical review may be sought from colleagues either during or after preparation of a manuscript. Advice from such sources is helpful, if not indispensable, in the formulation of ideas; but it is seldom sufficient in itself. Colleagues usually already know a good deal about an author's project from earlier conversations, and hence may fall into the same traps as an author. Moreover, colleagues' reviews are apt to be cursory. Review by coworkers, professors, or superiors, though highly recommended, should not replace objective review by outsiders.

Large research organizations that produce publications, as well as journals that publish formal papers submitted by the technical and scientific community, usually maintain a technical-review system for manuscripts. The purpose is to provide independent advice to an editor on the technical quality of submitted manuscripts, to aid in selection or rejection of manuscripts, and to help authors in improving their presentations. Technical review is considered an essential part of the writing-publishing process.

Manuscripts submitted for publication range from poor to excellent. Not surprisingly, the overall quality, both in thought and expression, is directly related to the quality and quantity of the technical review a manuscript has already undergone. Differences between manuscripts from independent geologists or small, understaffed and overworked college faculties and from larger institutions that can maintain elaborate systems for manuscript processing and in-house review can be great.

Typical review system

Technical-review systems are almost as numerous and varied as the publishing houses that have adopted them. The system described is reasonably comparable to those followed by many earth-science journals.

In this review system, the editor, usually a volunteer, is responsible for a society's entire publication program. The editor's chief responsibility is to make final decisions on the technical and scientific quality of all papers and books that the society publishes. No one geologist can know enough about all the earth sciences to make the required judgments. Instead, the editor relies heavily on a group of associate editors and on technical reviewers chosen by them.

When a new manuscript is received, the editor scans it just enough to decide which associate editor is most likely to be able to pass judgment. These associate editors, all volunteers, are chosen to provide as broad a spectrum as possible of topical and geographic knowledge of the specialties that are collectively known as earth science.

The associate editor is asked in advance, usually by phone, if he or she has the time and the requisite knowledge of the subject matter to deal with the new manuscript promptly. If so, the associate editor selects one or more technical reviewers who are broadly or specifically knowledgeable in the subject matter, and who are able and willing to review the manuscript for the society.

The associate editor and the technical reviewers are asked to read a manuscript and advise on its acceptability from the standpoint of scientific soundness, originality, breadth of interest, and length. In judging length, some compromise is usually desirable between the author's need to tell a complete story and the editor's need to conserve space.

Technical reviewers send recommendations to the associate editor as written commentaries or as marks and notes throughout the text. The associate editor reviews the manuscript and the reviewers' comments and advises the editor of the combined results. The editor must make the final judgment on publication — acceptance as is, acceptance upon revision, or rejection — and transmit it to the author. Reviewers may remain unknown to the author, unless they agree to have their names made known. The author may also be anonymous to reviewers. Many publishers believe that an anonymous review process provides more valuable criticism of manuscripts.

Organization and style also are examined by technical reviewers, because no research results, no matter how excellent, will be read or understood if they are poorly expressed. Matters of style and expression are considered less important at the review stage; they can be improved later while the manuscript is being polished and prepared for the printer. The chief objective of the technical-review process is to help the editor decide whether to accept a given manuscript, not to rule on details of its presentation.

After a review process

A review process is not intended to stifle an author scientifically nor to dictate preconceived notions held by technical reviewers. No reputable publisher wants to act as a scientific censor. Publishers seek scientific excellence, neatly and clearly presented. The purpose of a review process is to find such excellence and to help an author bring out the best in a manuscript for the ultimate advancement of science.

A tiny fraction of submitted manuscripts comes through the review gauntlet unscathed; these are recommended for publication virtually without change. They may be turned over to a managing editor, whose staff will prepare them for publication.

A number of manuscripts are rejected after they have been reviewed. Rejection may be based on judgments that the subject matter belongs more properly in some other publication. Other reasons may concern elements that reviewers were asked to consider, such as relevance and currency of research as well as adequacy of supporting data, references, cost of illustrations, and manuscript length. Editors realize that, no matter how firmly rejected, most manuscripts will inevitably come back to them or go to other journals. That is good, for almost every piece of research contains some elements of truth that deserve publication somewhere.

A vast majority of reviewed manuscripts are returned to authors with requests for revision, which range in scope from minor to major. A major revision may require what amounts to a completely new paper. In such cases, an editor may put the new version through the review process again, often using different reviewers from those who read the earlier version. For minor revisions, an editor may read enough of the revised version to be sure that an author has made a conscientious effort to follow reviewers' advice. If so, the manuscript is accepted for publication and put in the mill.

Author and reviewer

Authors react to criticism of their manuscripts in a variety of ways. Some welcome help, or at least acquiesce gracefully; others fly into a rage and see ignorance, if not foul play, in every mark made by technical reviewers.

Although authors seldom believe it until they become reviewers themselves, reviewers are usually people of good will, genuinely trying to help an author. Reviewing can be a thankless job. Rarely does a reviewer read a manuscript that contains some germ of new thought within his or her specialty. More often, a reviewer is an unpaid volunteer, doing a job out of loyalty to science and to the publishing society or institution.

Authors should approach reviewer comments with an open, cool mind. A reviewer is on their side, and every comment deserves thorough and objective consideration.

Fairness — and tact

Charges of prejudice, self-dealing, and other forms of foul play come quickly to the minds of a few authors when they receive their manuscripts back from a review process. Dealing with such charges fairly constitutes one of the more delicate responsibilities of a journal editor.

It is sound policy to accept charges of foul in good faith. An author should be offered another hearing by a new set of reviewers and should be given a chance to nominate some of the reviewers. Often authors are offered the opportunity to nominate reviewers in an initial review process. Regardless of how such charges are handled, an editor must be the final arbiter. If an author disagrees with a reviewer's comments and refuses to make revisions, an editor may request that the author respond to the review in a written rebuttal, which is published along with the article.

In cases where a reviewer points out plagiarism or duplicate submission of a manuscript, an editor should attempt to verify the charge and communicate with an author at the level of seriousness that journal policy and transgression dictate.

IMPROVING YOUR MANUSCRIPT

- Technical review is helpful — if not indispensable.

- The purpose of a review process is to bring out the best in a manuscript for the ultimate advancement of science.

CHAPTER 11

Editing and Proofreading

With the advent of word processing and electronic publishing, it is increasingly important that authors take responsibility for making manuscripts complete, accurate, and well written. You should submit the best copy you possibly can, making sure that it conforms to the style of the journal that will publish it.

Editing

Editing is intended to be flexible and to enhance an author's expression of scientific information. Style rules should not be so rigid that they cannot bend.

After a manuscript has gone through a review process, it is ready to be copy edited. Copy editors do the following tasks:

1. Identify and number all parts of an author's manuscript and artwork, such as illustrations, tables, and photographs.

2. Check for organization, grammar, and punctuation.

3. Make sure that a title is specific and concise and includes a locality, if appropriate; that an abstract is short, informative, and specific; and that the body of the text is concise and the style consistent.

4. Query an author on points of clarity, need to condense, and if necessary, suggest another choice of words.

5. Check cited references in the text against a reference list or bibliography to make sure no omissions or superfluous entries occur.

6. Mark a manuscript for style, such as italics and boldface.

7. Make sure that the order of headings and subheadings is logical and consistent.

8. Proportion artwork and check it for spelling, drafting, captions, and credits.

9. Make sure that all necessary permissions have been obtained and that they are properly worded in the text.

10. Return a manuscript to an author for final approval.

11. Reread a manuscript when it is returned to incorporate changes or additions.

Until the time comes when there is no editor between you and the reader, your responsibility ends with your approval of an edited manuscript. An editor marks a manuscript for a typesetter and marshals proofs through the printing process. Proofs are sheets of printed material that are checked against a manuscript and upon which corrections are made. At this stage, camera-ready copy has not yet been produced. Authors may be called on to read one or more proofs, but editors check author notes and relay changes to a printer.

Conclusive clarity

The first aim in editing manuscript copy or marking proofs is clarity. Clarity in this sense is used to mean making marks that everyone — author, editor, reviewers, and typesetters — can understand. Manuscript copy and proofs should be marked so clearly that the type can be set by a typesetter who knows no English — only the alphabet.

Manuscripts should be double or triple spaced so that a change can be inserted where it is needed. Manuscript copy should read in a continuous line, without distracting detours from the middle of a page to a margin and back again. Such detours invite typesetter errors. Margins should be reserved for instructions to a printer.

Each person who marks a manuscript or proofs should use a pencil — never a pen — that is a different color from that used by others. Initialing the copy will show the color that each person is using.

Copy editors correct author's work and prepare it for typesetting or word processing. Proofreaders correct a typist's or a typesetter's work. Copy editing marks and proofreading marks are generally the same. The difference is how they are shown. Copy editors work with double-spaced text, so they have room to mark changes in the text. Proofreaders work with galley or page proofs that have little space between lines. They mark changes twice: in the text and in the margin. The point of change is marked by a caret (^), and the correction or correction symbol is marked in the margin. When a line requires more than one change, marginal corrections are assembled in the proper order, separated by slash marks. If the number or juxtaposition of changes seems likely to be confusing, the best course is to kill the entire word or line and insert the proper form in the margin.

Generally accepted symbols are shown in the figure. A few common ones have been omitted because of duplication or ambiguity. (We see no reason to encourage use of, say, a half dozen variations of the deletion symbol.) In general, anything not to be set in type should be circled: for example, an editor's query to an author or instructions to a printer.

Proofreader Symbols

Authors, editors, and proofreaders use a language of standard symbols to indicate corrections to be made. Thus, a printer need not understand English to make the changes.

Mark the text to show a change, and write the appropriate symbol in the margin. A sample of marked copy follows these symbols.

Symbol	Meaning	Symbol	Meaning
⊙	insert period	(break)	break/Begin new line
⋏	insert comma	¶	paragraph
:/	insert colon	no ¶	no new paragraph
;/	insert semicolon	☐	indent 1 em
=/	insert hyphen	☐☐	indent 2 ems
ꜚ	insert apostrophe	(ital)	italics
ꜚꜚ/ꜚꜚ	insert quotation marks	(caps)	caps
(/)	insert parentheses	(c+sc)	caps and small caps
[/]	insert brackets	(l.c.)	lowercase
(set)?	insert question mark	(b.f.)	boldface
(set)!	insert exclamation point	]	move right
(shill)	insert virgule, slash, shilling	[	move left
en dash	insert en dash		down move
em dash	insert em dash		
⋏	subscript		move up
⋁	superscript	(straighten)	align horizontally
ℐ	delete 1 character	(align)	align vertically
ℐ	delete and close up		
a	insert from 1 character up to 7 words		center horizontally
(out, see copy, p.x.)	insert more than 7 words		center vertically
(tr)//	transpose adjacent letters		
(words or)		(stet)	ignore marked change
⌒	close up space	(sp)	spell out (abbrev.)
(eq.#)	equalize space	(score)	underscore
(#)	insert space	(rule)	use rule
(less#)	less space	(x)///	dirty or broken letter
(run on)	no new line	/?	query to author

Three-component triangular diagrams, introduced by the Finnish petrologist Petti Eskola, are widely used in petrology. ACF and AFM diagrams and modified version of these diagrams that are based on different components usually used to indicate what minerals are compatible with diverse metamorphic facies. Both pertain to silica-saturated parageneses. The afm diagram can also be used as just described but is more frequently used to show patterns of variation among rocks that appear to have been derived from the same consanguineous magmas. Triangular diagrams, in general, are used to show graphically chemical compositions that may be expressed in 3 components. The diagram below shows how the components are plotted. The numbers along AB indicate percentages of A, those along BC indicated percentages of B, and those along CA, percentages of C. In AFC diagram, C = Ca

F = (FeO + MgO + MnO)

In both of these cases, the oxide components are recalculated from the chemical analyses of the rocks to a 100 percent (molar) for the components plotted. Some petrologists modify the above outlined calculations by computing Fe as FeO, thus eliminating the Fe_2O_3 components from "A.

*Circled type indicates an instruction to the printer. Other letters and words will be typeset.

References

Bishop, Elna E., et al., 1978. *Suggestions To Authors, of the Reports of the United States Geological Survey*, 6th ed. U.S. Government Printing Office, Washington, D.C.

The University of Chicago, 1982. *The Chicago Manual of Style*, 13th ed. The University of Chicago Press, Chicago.

U.S. Government Printing Office, 1986. *A Manual of Style*. Gramercy Publishing Co., New York.

Proofreading

The best tool for an accurate proofreader is a skeptical mind. Mistakes cost money if they are repaired, and embarrassment if they are not. Editors and authors must each assume that no one else will read proofs; they alone are responsible. Authors may find it helpful to proof by reading one copy of a paper aloud to another person, who checks it against the original manuscript. Techniques to slow down your reading may make it easier to find errors. Some proofreaders read only one line at a time, covering the lines below with a ruler or paper. Others proofread a manuscript from back to front to concentrate on individual words rather than on the content of a paper. Read proof as if petting a porcupine: very, very carefully.

Most typesetting machines produce copy ready to be pasted down for printing by offset methods. Copy can be corrected by reinstructing the storage tape and "replaying" the entire copy, or corrections may go back to the operator, who either resets the lines required or produces words or letters to be used for correction. Authors and editors should send in as clear, unequivocal copy as possible and refrain from making "nice" but unnecessary changes.

Hyphen hassles

Words that break at the end of lines are hyphenated according to the word-processing program used. While many such programs are very sophisticated and are correct most of the time, others often break words in inappropriate places (such as *inco-rrect* and *prod-uce*). Mistakes are costly and troublesome and require a live person to deal with them.

In the days when only printers printed, the problem was not great, because they knew, as a part of their business, how to split words properly. Today, you may be your own typesetter. Basic principles for hyphenation follow:

o Try to avoid splits.

o Split words by syllables; consult a dictionary when in doubt.

o Distribute splits over a page so that they do not call attention to themselves (for example, four or more lines ending with a hyphen is unacceptable).

o Avoid splits at the end of pages or before tables or illustrations.

o Break words in logical places so that readers will not guess the finish incorrectly.

In geological work, you should mark a manuscript and check the proof to make sure a typesetter has not improperly run together rock and mineral terms or other technical phrases. For example, if the first half of a term such as *dihexagonal-dipyramidal* appears at the end of a line, it may very well show up as *dihexa-gonaldipyramidal.*

Permissions

Authors are responsible for obtaining permission from a copyright owner to use quoted material of more than a few paragraphs and artwork from previously published material. They should send a copy of all permissions to the editor. If you have not obtained permissions, check with the editor to find out about getting them. A written letter of permission must be on file before you publish anyone else's words or artwork — it is the law.

Copyright

You may want to copyright your own work. Under current copyright laws, a copyright notice consists of three parts: (1) either the symbol ©, the word *Copyright,* or the abbreviation *Copr.;* (2) the year of publication; and (3) the name of the copyright owner.

Many publications require that you sign over copyright of your technical articles to them prior to publication. Assigning them copyright is generally not a problem and makes it easier for other publishers who may want to reproduce your work to find a source for permission (authors move often and can be difficult to locate, whereas publishers are often more stable institutions). If a publisher owns the copyright for your work, you must obtain permission to quote large amounts of text from that work in another publication.

If you copyright your work, you may also want to register it with the Copyright Office in the Library of Congress. Registration is not necessary for copyright ownership. For more information, contact the Copyright Office.

CHAPTER 12

Editing in Style

Occasionally geologists find themselves editing other people's reports. This situation may be for a two- or three-year appointment or for an extended period. It may involve a small and rather informal publication appearing only a few times a year or an established monthly journal. To make your life easier as an editor, guidelines can be used to make style decisions.

Most publications have their own rules of style; some may have stylebooks. Since stylebooks are intended to cover all matters of editorial usage, they are often lengthy and difficult to use. They may be full of whims, such as a preference for *grey* rather than *gray*. But the basis of any style must be found in the subject matter of the particular publication involved.

Whims aside, no detailed stylebook can apply in all cases to all the various fields commonly lumped under "earth science." For example, paleontologists use the term *new species* so often that repeated spelling out with species names would be intolerable; but geochemists see the term so rarely that *n.sp.* should not be used in their journals.

Finding a stylebook

If no particular style has been ordained for a writer or editor, you should first examine stylebooks (if any) in your general field. For example, paleontologists may find that they can easily adapt the Council of Biology Editors style manual to their needs. As a news magazine, *Geotimes* uses the *Associated Press Stylebook*. Writer and editors should mark up a stylebook freely to change rules as time passes and evidence accumulates to show what is most effective for writers, editors, and readers.

Literary style, being less dependent than editorial style on subject matter, has been covered extensively in works on writing. If in doubt on this point, start with chapter 5 in *The Elements of Style* and take it from there. Also, if you are a specialist who deals in writing and editing, you should assemble your own collection of clichés and examples of careless usage and inept syntax, taking care to avoid the merely arbitrary rules so common in such lists.

Rules for typographic style — that is, the style and size of type — are usually fixed for a given periodical. A pure example of such a style sheet is hard to find in published form, for they are seldom useful to anyone but editors of the publications involved.

Codifying style

If you are a new editor and find that your publication's typographic style has not been set down in writing, start right away to set it down if only as a convenience to your successor. Confer with the typesetter and printer, which will also greatly improve your communication with them and add to your understanding of the publication. A written typographic style is essential when the time comes to call for typesetting and printing bids.

A common but primitive approach is to leave typographic style decisions up to a printer, who will follow the style that was used for previous issues. That course virtually bars improvements, for when you try to change a style point, a printer is likely to consider the change an error. Worse, the practice encourages a printer to believe that you do not know your own business, which may be right.

For a new or one-time publication, style conventions often seesaw. As an editor, you may decide that a journal should consistently use the form *per cent*, but after publishing a few papers with statistics requiring frequent use of the term, you change the style to the symbol %. Later, you may find that statistical papers have become exceptions and that *per cent* (or perhaps *percent*) was more fitting after all.

In working on *Geowriting*, contributors used a variety of styles. One contributor argued for a general policy of Latin forms rather than Anglicized versions (for example, *symposia* rather than *symposiums*), another for the shorter form of a word that has two spellings (for example, *dialog* rather than *dialogue*). As an aid to determining style, we editors made up a short list of style questions as we went along — a list that others may find useful:

- o Capitalization or lowercase. National Geographic Society vs. National Geographic society; the Editor or the editor.

- o Italics or quotation marks. "Modern English Usage" or *Modern English Usage*, for titles of books, periodicals, and chapters.

- o Quotation marks. 'English' or "American".

- o Commas, periods, and quotation marks. It is common practice to place commas inside quotation marks, as in "The New York Times," in all cases, but some prefer to place them outside of the quotation marks when the comma is not part of the name or quotation: "The New York Times", for example.

o Abbreviations. Sacramento, California, or Sacramento, Calif. (or Cal., Ca., or CA); USA or U.S.A.; No.1, no.1, n.1; 3 kg or 3 kg.; per cent, percent, %.

o Numbers. Use numerals for measurements; in text spell out for one through nine, use numerals for 10 and above: eight, nine, 10, or 8, 9, 10; 14 6-point slugs or 14 six-point slugs; 1000 or 1,000; 28,000,000 or 28 million, or 28x106.

o Hyphens. For example, cooperate, co-operate.

o Latin or English. Curricula, curriculums; formulae, formulas; symposia, symposiums.

o English units or metric (International System of Units) or both.

o Apostrophes. 1970's, 1970s.

o Long form or short. Employee or employe, gauge or gage, catalogue or catalog.

o Variant spellings. Gray or grey, skilful or skillful, dialogue or dialog.

o 1 word or 2. Artwork or art work; stylebook or style book; halftone or half tone; checklist or check list; groundwater or ground water.

o Citation style. Word order, abbreviations, italics.

o Paragraph indentations or block paragraphs.

o Special typefaces.

o Inclusion of sections of a manuscript as well as the style in which they are constructed: Glossary? Index? Checklists? Appendices? Bibliography? Citations?

As you work on a publication, consider the following categories in making style decisions:

o Capitalization. For example, *Earth's mantle* or *earth's mantle.*

o Punctuation. For example, use of the serial comma; *U.S.G.S.* or *USGS; 1990's* or *1990s.*

o Compound words. For example, *art work* or *artwork; ground water, groundwater, or ground-water.*

o Numbers. Spell out or use figures (for example, spell out one to nine and use figures for 10 on); *1000* or *1,000;* 28,000,000 or *28 million;* English or metric measurement.

o Spelling. American or British (for example, *color* or *colour*); variant spellings (for example, use the first listed form: *archaeology* rather than *archeology*).

o Abbreviations and acronyms. For example, spell out *United States Geological Survey* for first usage in text, then use abbreviation for subsequent references; *3 kilograms, 3 kg.*, or *3 kg* (without a period).

o Organizational style for parts of a publication, such as for a glossary or an index; and for parts of a chapter, such as style for paragraphs, running heads, and subheads.

o Typestyle. For example, italicize Latin names and publication titles.

CHAPTER 13

Writing Reviews

Reviews of books for the general public are written by professionals who live by their trade. Those people cover creative works, such as novels and poetry. The reviews reflect the taste or intellectual and social judgment of an individual who knows that readers want a personal reaction as a helpful guide in choosing what to read or buy. General book reviewing is not our model here.

We speak of reviewing scholarly books as a means of education in a broad and disciplined sense. These are books that purport to add something to the body of human knowledge, books written for serious students and scholars, books that purport to be original contributions in analysis or in form, emphasis, and content. A quite different responsibility rests on scholarly reviewers.

When you are a reviewer, you are not a professional writer; rather you are a specialist in a certain field of knowledge. You are chosen for that knowledge, for your critical awareness of the work and thinking going on in the subject, and for your sensitivity to advances in a field. Specialists may be inexperienced at review writing; indeed, book review sections of some journals reflect that state.

You should remember, as your first commandment, that you are responsible to your readers, who are mostly fellow professionals, students in the broadest sense. You are not responsible to an author, who should be prepared to accept honest, objective, and competent evaluation of his or her work. Nor are you responsible to an editor, board, or group that chose you. Unlike a general book reviewer, who offers a subjective opinion, you are required to subordinate your personal feelings to objective appraisal.

What should a specialist review tell its readers? The vital statistics — title, author, number of pages, date of publication, name and location of publisher, and price — are obvious.

You should read a book's preface and any introductory statement to be aware of an author's intentions as to readers and purpose. Otherwise, you are apt to play expert and say how you would have written a book on the subject. Reviewers should not criticize authors simply because the author did not write the kind of book the reviewer wanted to see written. Rather, reviewers should judge how effectively the author accomplished their articulated purpose.

First, what reader is addressed? Obviously, *An Introduction to College Geology* as a title speaks for itself. *Cranial Muscles of* Lucanus cervus *in Iceland*, however, will require some explanation. Review readers need to know a book's scope and its level of technicality. If a beginning geology book assumes a background in calculus and physics or in comparative anatomy, that information is relevant in a review because it restricts the readership.

Second, what is the original value of a book's content? What new knowledge or what new emphasis or insight does a work offer? In what respects and to what degree is a book unique? Comparisons with other published works in the same field of knowledge may be helpful.

Third, how successfully does an author execute a book's purpose? Is the information or analysis presented logically, lucidly, and consistently? If weak in any such area, is a book still worth the effort of study?

About lucidity: As emphasis on rapid and frequent publication has grown, many authors have slipped into bad habits. There has been haste to grind out writing, with no respect for lucidity of organization and exposition. One could describe many of the results as written in "academese." Lucidity is still the hallmark of a good book.

About accuracy — and errors, small and large: Small errors, if prevalent, call for comment. Yet, statistically, possibilities for errors per page are in the millions. Small errors are not usually worth much comment. A reviewer should evaluate the validity of the content in general — the large matters.

After you have completed the evaluative tasks, you have the right and duty to express — parenthetically, perhaps, — your personal reactions to a work. Yet you must observe a fine line. You must be objective. You may show favor or regret, enthusiasm or distress. But you should, in any event, be reasonable, not subjectively personal, always remembering that your primary responsibility is to readers.

A specialist reviewer is often chosen by a board or an editor, with the idea that a specialist's work and a book's content have a professional affinity. The person chosen usually feels a duty to follow through and write a review. But sometimes, acting on this duty may not be valid. If a first reading indicates to you that you are not suited to the task, you will best serve your fellow professionals and yourself by withdrawing.

Specialty reviewing is an underdeveloped art because of the responsibilities of reviewers, who are not usually chosen for their writing talent. Yet, do the best you can.

Any review, favorable or not, is publicity and usually results in sales. Good reviews may do more than help readers choose before reading. When your reviews help readers avoid bad books, make them aware of reasonably good ones, and discover those that are singularly good, you are also helping to achieve higher standards and (let's hope) the making of fewer and better books. The practice of this underdeveloped art may thus become a rewarding responsibility.

WRITING REVIEWS

A book review should objectively evaluate a book's

- Stated purpose.
- Intended readership.
- Scope and level of knowledge.
- Original value of content.
- Lucidity and accuracy.

CHAPTER 14

Writing for the Media

Newswriting is so different from other types of writing that it is probably inefficient for most scientists to write nontechnical descriptions of their work. Many scientific organizations employ a person to handle the writing and distribution of announcements to the media — newspapers, magazines, television, and radio. But in small organizations or societies, scientists are occasionally asked to write news stories about their work. Scientists who find themselves in that position should obey some simple rules of newswriting to make the process easier.

Journalists are nearly always on very short deadlines. For radio and television reporters, such deadlines are daily or even hourly; for newspaper reporters, daily and weekly. Magazine writers have the luxury of weekly or monthly deadlines. In most cases, journalists work in shorter time frames than scientists. Because scientists and journalists work in such different worlds, they may have some difficulty working together. When writing for the popular media, scientists must adopt a journalist's role and undertake writing a journalist's way.

Newswriting is different from writing used for technical communication and from most other kinds of expository writing. It requires much less technical language and a different selection of facts. The most notable difference is in the order of information. In contrast to a technical article, which leads with an abstract or an introduction to a work, a news story requires that the most important information come first, usually in the form of a one-sentence summation of a work's results. Newspapers have a limited amount of space; when editors cut stories to make them fit that space, they generally start slicing from the bottom. Information at the end of a story may not make it into a newspaper. Also, newspaper readers and broadcast listeners are accustomed to getting important information quickly. Thus, as a journalist, you need to select the most significant information and put it first. That information is generally the results of your work and their consequences and significance, the sort of thing that waits until the very end of a technical article.

Journalists call the first sentence the lead, and it is the most important part of any news story. While journalistic practice once was to answer five "W" questions — Who? What? Where? When? Why? — in a lead, today's leads are generally less all-encompassing. News leads, however, should include such information as the source of a story — that is, the person who wrote the article or made the statement that is the source of the news in a news story — where the person works, and either the person's official title or specialty. Newspaper editors and newspaper readers often judge the credibility of a news story on the basis of the person who is the source of the information, and you should supply the source so that both can make that judgment.

Generally try to avoid using *Dr.* in a person's title unless you are referring to a physician. When a story is based on a publication that has three or fewer authors, list them all. For any number greater than three, list the first author, followed by a phrase such as *and colleagues;* for example, *according to Dr. C.C. Chang and colleagues at the University of Kentucky.*

Steps for writing a news release

1. Write the first page on letterhead paper that supplies the name, address, and phone number of a story's source and the date of the news release.

2. Use a headline, a line that describes in a few words the essence of a story.

3. State a story's location before a lead sentence.

4. Begin a story with a lead sentence.

5. Add more specific information after a lead sentence.

A lead sentence is the most important part of any news story because it is the sentence that people are most likely to read. What's more, if well written, the lead may pull readers into the rest of a story. Even the first word in the sentence is important. Virtually any science news story could start with the word *Scientists: Scientists at the University of Wisconsin today announced that they had isolated the gene responsible for some forms of muscular dystrophy.* While scientists are the source of the news, readers are more interested in the news itself. Thus, a better lead would start with the subject of the release, which is muscular dystrophy: *Muscular dystrophy is another step closer to being understood.* Try to avoid leads that begin with the word *the* or any other ordinary word. Try to avoid question leads, such as *Will scientists ever be able to cure muscular dystrophy?* The purpose of leads is to answer questions, not ask them. (See examples on pages 67-68.)

In the first paragraph or two, concentrate on the significance of the finding, not the methods of research. Readers of nontechnical articles are generally not interested in methods; they have no intention of trying to replicate anyone's work. They assume the work is credible because you have chosen to write about it and the newspaper has chosen to carry the article. In some cases, late in a story, it may be pertinent to include a paragraph or two about methods; sample sizes may be particularly appropriate. Unless a method is the story, however (such as the development of a new scientific technique), methods usually require little discussion.

In addition to explaining a story's significance and source, you should supply other information that may interest readers. Strive to anticipate questions readers may have, and then answer them. Always include the source of funding for the research; that information is crucial in helping editors and readers determine a story's credibility. Feel free to use quotations from a source. That person can provide alternative explanations of work, usually in less technical language. Quotations give stories a personal feel that is important in writing about science. If a story has a local or regional angle, feel free to point it out. Community newspapers tend to be less interested in national affairs than in what's going on in their backyard. If you discuss location, your story is more likely to be used. For example, location information is particularly important when writing about upcoming field work. Also, keep in mind that names make news; a newspaper will often use a story simply because a source is a resident of its community.

SUGGESTIONS FOR NEWS RELEASES

- Use nontechnical words and phrases. If you must use technical terms, define them. Never assume your audience knows a word's meaning.
- Use short sentences and short paragraphs.
- Include a source's title and affiliation.
- Use quotations in plain language.
- Keep stories short.

Write stories as long as they need to be, but keep in mind that newspapers have limited space. If possible, keep news releases to one double-spaced page. Announcements of grants or upcoming meetings, for example, can usually be held to one page. Limiting a story's length will dramatically increase the chances that newspapers will use a story. Long pieces not only take more space in a paper, they also require more time to enter into a newspaper's typesetting computers. By limiting information to the smallest space necessary, you make it less likely that an editor will need to cut your story and thus remove information that you would like included.

Try to stick with short sentences and short paragraphs. Research has shown that shorter sentences are generally easier to understand than longer ones, although varying sentence length is still a good idea in any kind of writing. Paragraphs in newspapers must be short because of the narrow columns used in newspaper formats. Numerous paragraphs create indentations, which make it visually easier for readers to keep their place in a column. Long paragraphs produce unbroken columns of copy that can be tough to read. Thus, newspaper copy is often made up of paragraphs that are only one or two sentences long. Do not try to follow an old rule from English composition that says every paragraph needs a topic sentence followed by several supporting sentences. In newswriting, sentences written in a readable order are far more important; thus, paragraph breaks may appear almost randomly.

Writing news releases for broadcast outlets, such as radio and television, is not dramatically different from writing for newspapers, although more emphasis may be on writing short, tight stories. If such stories are broadcast, they will often involve interviews, either over the phone or in person, so that audio tape (called an "actuality") or videotape can accompany a story to make it more interesting.

In sending news releases to the media, try to keep a couple of factors in mind. First, make sure a source is available to answer any questions a reporter may have. Don't send a release out the day before a source is preparing to leave for three months of field work in China. Also, try to avoid sending releases during times when other news will push them out of the paper, such as the week before an election. The period between Christmas and New Year's Day is often a slow news time and thus a good time for a story to get prominent attention. Try to avoid mailing a story when some other breaking story is dominating the news.

If you are not the source of a story but are writing a news release about someone else's work, a question of clearing the story with the source will arise. Similarly, if you are the source of a story and send a release to a newspaper, which leads to an interview, a question of checking the final story may come up. Science stories often have subtle and important implications that you may not recognize. Changing one word sometimes makes the difference between an accurate and an inaccurate story. When doing releases, taking the time necessary for this final check is a good idea. But don't expect the same treatment from reporters. Most newspapers and magazines have a policy against such checking; even if no such policy existed, time often does not allow for checking. Some reporters may call to read relevant portions of a story, or parts they have questions about, and ask you about their technical accuracy. While they welcome technical correction, they are generally not looking for help with the style of their presentation.

You may be asked to write longer feature stories for the press, stories that are not based on breaking news but allow more in-depth study of a research project or scientist. Such stories are often difficult to write, however, and for the most part, reporters prefer to write them.

Ways to increase the chances that your news release will go into print instead of on an editor's kill spike:

- Double-space all copy so an editor can mark it easily and clearly.

- Do not underline; few newspapers use italics.

- If a release is longer than one page, type <u>more</u> below the end of the last line on all pages except the last.

- Avoid breaking words from line to line, and never break technical words (or you may find an unintended hyphen in print).

- Do not break a paragraph at the bottom of a page.

- Use an end mark: <u>end</u> or <u>30</u> or <u>###.</u>

- Staple all pages together.

- If figures are included, inform an editor after an end mark, not in the text (or your instructions may inadvertently be set in type). Write captions for figures.

- Proofread the final copy. Do not ask a typist to proofread it. Ask a third person to read it, too. As an extra precaution, read it aloud.

- Ask yourself, Have I written the news for <u>readers?</u> Have I made it easy for an editor to use?

The following samples illustrate two new release formats. Also note headlines, lead sentences, order of specific information, and use of quotations.

Washington University
University Communications
Campus Box 1070
One Brookings Drive
St. Louis, Missouri 63130-4899

Judith Jasper
Executive Director
Office: (314) 935-5230
Home: (314) 822-2425
Fax: (314) 935-4259

News Release

Contact:

Debby Aronson
(314) 935-5251

EARLIEST COMPLETE SKULL OF PRIMATE ANCESTOR FOUND IN EGYPT

St. Louis, Mo., November 15, 1994— A Washington University anthropologist has found the complete skull of a 35-million-year old prosimian in the Fayum region of Egypt.

This is the earliest fossil evidence for modern prosimians, one of the three evolutionary lineages for primates.

"There have been a large number of archaic prosimian fossil finds, but the origins of the modern prosimians have been a mystery," said D. Tab Rasmussen, one of the scientists that analyzed the skull. "This find proves that the evolutionary group that gave rise to modern prosimians was present in the Fayum 35 million years ago."

Fossil evidence of the other two lineages, anthropoids and tarsiers, already had been found in the same region. This makes Egypt the only place in the world to have evidence of all three branches of primates. This suggests that all three primate groups evolved in Africa and that their common ancestor, the creature that links humans and all other primates, may yet be found in this region, Rasmussen said.

Rasmussen, Ph.D., associate professor of anthropology, excavated the complete skull in the Fayum, an area north of Cairo and west of the Nile River renowned for its fertile soil and extensive archaeological resources, fall of 1993 with his colleague Elwyn Simons, Ph.D., James B. Duke Professor of Anthropology at Duke University. Their findings were published in the recent issue of the Proceedings of the National Academy of Science.

Primates are divided into three groups: anthropoids, prosimians and tarsiers. Anthropoids include humans, monkeys and apes; prosimians include lemurs, lorises and bushbabies; and tarsiers, of which only one type survives, and is called a tarsier. Tarsiers, which are small, nocturnal, tree-living animals living in parts of Southeast Asia, like Borneo and the Philippines, share characteristics of both anthropoids and prosimians.

Prosimians have open eye sockets, unlike anthropoids, which have closed eye sockets. Other characteristic features of prosimians include a "tooth comb," so called because the canines and incisors in the lower jaw jut straight out and form a comb the animals use to comb each other's fur. In addition, prosimian canines in the upper jaw are flat, like daggers, rather than cylindrical, like candy corns. Anthropoids have cylindrical-shaped canines.

Not only is this the earliest fossil skull of the prosimian, it is the first fossilized evidence of the tooth comb. One tooth is preserved in the lower jaw of Rasmussen's specimen.

-over-

67

Stanford News

Stanford University News Service
Press Courtyard, Santa Teresa Street
Stanford, California 94305-2245
Tel: (415) 723-2558
Fax: (415) 725-0247

12/13/94   CONTACT: Janet Basu, News Service (415) 641-7198

Seismic 'lending library' helps geophysicists profile the earth

STANFORD — Lining the wall of a sub-basement room in Stanford's Mitchell Earth Sciences Building are more than 250 toaster-sized gray boxes, each costing as much as an economy car. Many of the boxes have traveled hundreds of thousands of miles, and they're familiar with the dangers of international travel: Some have been mauled by bears, one was shot, one was pried open by an African tribesman.

These are seismic recording devices, critical for studying geologic structures deep beneath the earth's surface. Called RefTeks, after the company that manufactures them, they are used to identify buried rock layers, revealing underground faults or making an image of the epicenter of an earthquake. Much as a CAT scan reveals the brain, they can help find surprises within the earth's crust – such as the mysterious horizontal slab of rock that has been detected 10 miles beneath the Bay Area, a possible link between the region's major earthquake faults.

"These instruments are for geophysicists what a telescope is for an astronomer," said Marcos Alvarez, one of three scientific engineers who manage a Stanford-based lending library for the seismic detectors.

Alvarez and his colleagues, Steve Michnick and Bill Koperwhats, act as lending librarians and technical support wizards for geophysicists from all over the world, who line up to borrow the RefTeks in batches of several dozen to several hundred at a time.

In the past three years, the equipment from Stanford has been used in 30 scientific projects, from Antarctica to the Alaskan Arctic. In any given week, a batch of devices that just cleared customs in Kenya may be headed next for the Northridge earthquake region in Los Angeles or for the Tibetan plateau. And at last week's American Geophysical Union meeting in San Francisco, more than 100 scientific papers in a dozen sessions reported on data collected with RefTek detectors and related instruments from Stanford.

Funded by a recently renewed $1.3 million grant from the National Science Foundation, the Stanford lending library is a PASSCAL center – part of the Program for Array Seismic Studies of the Continental Lithosphere. With its sister center at Columbia University's Lamont Doherty Earth Observatory in Palisades, N.Y., it is sponsored by a multinational consortium of universities called IRIS – Incorporated Research Institutions for Seismology. Membership in IRIS costs each institution $2,500 per year and gives its researchers free access to the equipment at PASSCAL centers.

Support for the best ideas

Why have a lending library for portable seismographs? The first reason is that these hardy but sophisticated devices cost about $10,000 apiece. A single seismology experiment can easily require 50 to 500 RefTeks – plus sensors, batteries and other accompanying equipment, and computer workstations to analyze the data. That quickly adds up to a cost beyond the reach of any one institution, Alvarez said.

(More – passcal)

68

CHAPTER 15

Design with Type

Use of type in design is both an art and a science. A logical approach is to treat typography first as a science and learn the nature of the thing and the mechanical rules that control its use. Out of that knowledge should come an understanding of the gray area between the black and white of the "rules"; for somewhere in that area the science becomes a craft and then an art.

Typography uses two basic units of measurement: points and picas. Type is measured in points; line length is measured in picas. Points are converted into picas except when specifying type size, which is always given in points. Equivalent measurements follow:

1 pica =	12 points
1 point =	1/72 of an inch
1 inch =	6 picas (72 points)

Point sizes range from 5 points to 72 points. In general, text type refers to type up to 14 points; display type is over 14 points. Display type is used to attract attention, for example, in chapter headings and advertising headlines.

Leading is the amount of space between lines. Proper leading adds to legibility by helping a reader's eye stay on a line. Too much leading wastes space and produces the impression that the type is for a child's eye. Leading also contributes to type color or texture, a characteristic you can see best by placing several publications on a table and standing just far enough from them that you cannot read the text.

Typeface refers to the specific design of an alphabet. Each typeface has a name. For example, Times Roman (originally designed for the *Times of London*) and Helvetica are two different typefaces. Times Roman is a serif typeface; that is, it has short cross-strokes at the end of some of the characters. Helvetica is a sans serif typeface; that is, no characters have cross-strokes. The differences between typefaces may be quite small, but it is worth noting that the lowercase *g* is often the most distinctive character. Other letters likely to be distinctive are *a, e, p,* and *t*. A font is a complete set of all the characters (uppercase and lowercase letters, figures, and punctuation marks) of one size of one typeface.

Typefaces can have a variety of typestyles. Times Roman belongs to what is known as the roman typestyle, which has serifs and graduated thickness of strokes. Some other typestyles are italic, bold, and small caps.

The best design is done by exploiting the possibilities of a limited number of typefaces and typestyles. Type specimen books show a large and attractive variety of typefaces and typestyles. Such variety often intoxicates a beginner, who may be tempted to spend too much time and money on too many typefaces and typestyles. Simplicity is best. Word-processing packages and page-layout programs typically contain the most commonly used serif and sans serif typefaces.

An example of how typeface, typestyle, font, and leading fit together is seen in this book. The main text is set in the 12-point Times Roman font that is available in Ventura Publisher. Other fonts used are **12-point Times Roman Bold** and *12-point Times Roman Italics*.

Design elements

With the proliferation of computers and desktop-publishing programs, you may be responsible for the design of your publication, especially if you are the editor of a society journal. Resource books are available with information on publication design and readability (please see section, Reference Shelf). You can make your publication more readable and pleasant to look at by using some general criteria of good design:

o Uppercase and lowercase letters are easier to read than all caps or all lowercase letters in titles, tables, and captions.

o Serif type is easier to read than sans serif type.

o Two or three columns of type on a page are easier to work with for placement of tables and figures.

o Two or three columns are easier to read than one wide column.

o 10-point type is attractive and legible in a two-column format; 9-point type is acceptable for a three-column format.

o Avoid excessive use of italics and boldface; do not underline.

Times	g a e p t
Palatino	g a e p t
New Century Schoolbook	g a e p t
Helvetica	g a e p t

CHAPTER 16

From Ink to Paper

Understanding something of the mechanics of the various ways of putting ink on paper helps writers and editors choose the optimum combination of methods, equipment, cost, and timing. It also helps prevent trouble for others.

Production route

The process of typesetting and printing usually follows these stages:

1. Editor sends marked copy to a typesetter.
2. Typesetter sets the type and pulls galley proofs, or an editor or word processor enters typesetting commands or formatting on a computer and prints out proofs.
3. Editor (and sometimes author) reads and corrects proofs.
4. Editor (or designer) cuts up galleys and pastes them up on artboard, producing a detailed dummy that shows exactly where each line of type and each illustration is to appear. In desktop publishing, an editor or designer uses a page-layout program to lay out both text and illustrations, thus producing an electronic camera-ready copy.
5. Using a dummy as a guide, a printer assembles the type into page forms and pulls page proofs. Or an artist cuts up a photocopy and fastens it in place. In desktop publishing, after final adjustments to a layout are made, the job is run out on a laser printer or sent electronically or on diskette to a printer or a service bureau. A printer or service bureau produces negatives on a Linotron or other electronic typesetting device and then prints out page proofs.
6. Editor checks page proofs, electrostatic copies, photocopy pages, the actual pages, or a blueline.
7. Printer makes plates and prints the job.

Photocomposition

With the advent of computer technology, many companies developed machines for high-speed typesetting, in which a computer generates 15,000 characters a second.

Proofs and proofreading are affected by the different systems of setting type. Proofs usually take the form of an electrostatic copy. Any change in an original means that someone must carefully cut out a part of a photo-copy and strip in a correction. In short, author's alterations weigh heavily against the high-speed systems and greatly increase the cost and the time — and the need for careful copyediting.

Do it yourself

Today, we are in the age of computers and video-display terminals. In many systems, a writer types the story on a keyboard and need never see the story on paper. Instead, the story appears on a video-display terminal. A writer can scroll a story up and down, make changes, delete characters, words, or paragraphs — or add them. A writer can ask the machine to do such things as search all occurrences of *adn* and then change them all to *and*. A writer can put on an editor's hat and command the machine to show how the story will appear if set in 9-point Times with 11-point leading, set 13 picas wide, with all hyphenation completed. A writer can say, "Sorry, I meant to have you set that 14 picas wide"; almost instantly a new version appears on the screen.

Many magazines and newspapers make up their pages on a screen and transmit them via satellite to printing plants throughout the United States, and the pages include artwork as well as type.

Platemaking

For centuries, original type was printed by letterpress: Images with a raised surface were inked, then impressed on paper. Beginning with the Linotype late in the 19th century, printers used temporary (remeltable metal) type. Later, in a variation that became almost universal, printing plates were made by way of molds from the assembled hot type. Letter-press was displaced by offset printing, a method that uses an intermediate medium (three cylinders) to transfer an image onto paper. Etched thin metal plates are made using photographic methods. The plates are wrapped on the first cylinder during the printing process. Making dupli-cate plates is common, especially by newspapers, so that several press units can print the same copy at the same time.

Such photomechanical plates were once used primarily to reproduce artwork. The simplest kind of artwork is a line drawing, which is anything consisting solely of black and white, with no intermediate shadings of gray. Ordinary type is line art. To make a photomechanical plate from line art, printers use a film negative of line copy (which may include type) and project the image onto a metal plate with a photosensitive coating. They treat the plate with chemicals that etch away the metal in the areas that are not to be printed.

Screens

A continuous-tone image (such as a photograph) has to be converted into line copy (dots) to be printed. Such copy is produced by placing a fine-line screen between a camera lens and the film carrying the image and photographing the film. The resultant film is developed to produce a negative. A plate is made from the negative. A printed image is called a halftone.

In the screening process, continuous-tone copy is converted into evenly spaced dots of varying size, shape, and number. In printing, the density and size of the dots resemble the gradations of tone in the original continuous-tone artwork.

Screens are measured by the number of lines per inch. The more lines per inch a screen has, the finer the dot pattern and the better the quality of halftones. The size of a screen partly depends on the paper to be printed. Most newsprint is a rough paper, so it requires a coarse screen (55 to 85 lines per inch) to keep space between dots from being plugged with ink. Smoother papers may permit a 150-line screen or higher, resulting in finer detail.

A printed halftone usually cannot be rephotographed for subsequent reproduction. When a new plate is made, the already screened halftone would be rescreened. The result combines the two screens and can cause a loss of resolution and a moiré effect, an undesirable pattern. An original screened negative can be enlarged or reduced only slightly. Normally, linework and halftone work should not be photographed together on the same negative, because screening breaks up lines into fuzzy rows of dots. They should be photographed separately and then reassembled.

Offset lithography and letterpress

In much the same way that computer generation has become dominant in typesetting, offset has taken over in printing. On most presses using the offset principle, a printing plate is wrapped around an impression cylinder, which transfers an image to a rubber blanket; the blanket, in turn, transfers — offsets — the image to paper. Hence, the common short term *offset*.

To make a lithographic plate, a printing image is projected onto a metal plate with a photosensitive coating. The plate is chemically treated so that the printing area is water repellent and the nonprinting area is oil repellent. Thus, in printing, the printing area will reject a water solution and accept ink, and the nonprinting area will accept water and reject ink.

Letterpress plates are also made photographically. The nonprinting areas are etched away until only the printing areas are raised and ready to be printed.

Ink

Now briefly, consider inks. No one kind suffices for all kinds of work. It is worth noting that the choice of inks is by no means limited to the standard colors shown in manufacturers' catalogs; almost any color can be matched and custom blended at a cost not prohibitively greater than black.

Previously, most printers used petroleum-based inks, but in recent years more and more have been willing to experiment with other types of ink, most notably soy-oil based ink. Soy ink provides excellent quality and color, keeps presses running smoothly and more cleanly, and does less harm to the environment because it is nontoxic and does not require toxic cleansing agents on the presses. Soy ink also is produced from a renewable resource, unlike petroleum-based ink.

Imposition

The question of inks comes up when black and one or more additional inks are to be printed on the same page. Extra inks can open up an enormous range of possibilities, but to cut costs and to simplify matters, remember that each ink requires a separate plate. If the number of pages is so large that more than one plate is needed to print only one ink, it may be possible to arrange the layout of pages (the imposition scheme) so that fewer plates are needed for the second ink. Imposition consists of assembling and arranging all the pages that are to be printed on one large sheet of paper. It must be done in such a way that after the sheet is printed and folded, the pages will appear in the correct order. A full sheet normally prints in units of 4, 8, 16, and 32 pages. When folded, these pages are called a signature.

As a few minutes with pencil and paper will show, even a booklet of only a few pages can be arranged in several imposition schemes, and the number of possible schemes goes up sharply as the number of pages increases. The most efficient scheme depends on factors such as the press and folding machine to be used, the number of pages on the sheet, the desired arrangement of signatures, and the characteristics of the paper.

Editor, designer, printer, and binder must all agree on the best imposition scheme for a job. Efficiency of binding is often the most important factor, but a designer may have an overriding need for, say, a second ink on certain pages — perhaps a map that requires color for clarity — at the lowest possible cost. One imposition scheme may allow a second ink on only consecutive pages in the middle of a signature. Another scheme may allow the second ink on the pages near the beginning and the end of a signature.

If a second ink is required on certain pages of a publication, the position of the pages will suggest an imposition scheme. That scheme will usually allow use of the same ink on certain other pages at little or no extra cost.

Paper

Quite possibly the most complex single factor in printing is paper. An editor should learn as much as possible and also call on an expert to help make most paper choices.

Paper is sold by weight, and weight is generally given as the weight per ream (500 sheets). Different papers come in different sizes, such as 25 inches by 38 inches for book papers and 20 inches by 26 inches for cover papers.

The best paper depends largely on the printing process that will be used. For example, many old and still popular typefaces yield their best appearance when printed by letterpress on a soft paper. Finely screened halftones may lose detail if printed on any other than a smooth, hard finish, but the same paper may produce too much glare for text.

More publishers than ever are using recycled papers in their printing. The quality of these papers is excellent and the variety seemingly endless in both cover and text stock. A printer can provide you with samples from paper manufacturers.

Binding

Folding and binding, falling as they do near the last stages of print production, are often unduly neglected. Much trouble can be saved merely by providing a printer early in the game with a blank dummy of the finished work, folded and trimmed to the exact size visualized.

A blank dummy can also be used to check sizes. A dummy should fit a standard envelope or box; otherwise, special containers will have to be made. Check and decide at this point, not after the job is printed, bound, and delivered.

Separate maps, included in the back of a publication or issued in an accompanying container, may be folded by a printer in such a way that they are difficult to use and nearly impossible to refold. Give a printer a sample of the way you want maps folded. Neat accordion-folds not only make a map a great deal easier to use but also increase its life expectancy.

Binding methods include the following:

o Saddle stitching. Usually wire staples through a centerfold.

o Side stitching. Uses an ordinary office stapler.

o Side-wire stitching. For thicker books, two wire staples placed 1/4 inch from binding. Method has given way for most part to perfect binding, so book can be opened flat.

o Perfect or threadless binding. Folded sections, or signatures, are gathered, the back fold is trimmed square, and the individual sheets are glued by their back edges to the cover.

o Casebinding or edition binding. The binding is Smyth-sewn (section-sewn) or side-sewn. In Smyth-sewn bindings, folded sections are saddle-sewn with thread, and individual sections are then sewn together. Side-sewn books have the stitching pass through the entire book 1/8 inch from the gutter, rather than have individual sections sewn. Thick paper-bound books can be Smyth-sewn and then perfect bound. Case or edition binding connotes hard-bound books. The cover is made separately and consists of rigid or flexible boards covered with cloth, paper, or other material, which surrounds the outside of the board.

CHAPTER 17

Looking Ahead

Most writers and editors work with computer-linked keyboards and screens rather than with typewriters, and with computer-stored data as well as traditional books and journals. Many "typescripts" are on magnetic tapes or disks as well as on paper. Technology will force writers to follow standards that so far have remained elusive.

Typescripts are often read by machine and displayed on a screen for checking and editing; type specifications are added, actual type is produced within minutes, and traditional ink on paper can easily follow within hours or even minutes. As computers condense the time between an author's keyboard and the reader's book or screen, they may increase the chance of error, neglect, and bad writing. It is up to the editor to see that they do not.

Extensive networks now link computer terminals and a variety of data bases. Researchers, writers, editors, and publishers are exploiting the ramifications. Publishers are assembling customized textbooks from material in data bases, adding new material and editing as necessary, and printing the books with press runs as low as a few hundred.

Microcomputer programs are available for word processing and typesetting, graphics production, and page layout for both Macintosh and pc-based systems. Other programs are available for checking grammar, writing, style, etc., although these should not be relied upon totally. If you have typed "from" instead of "form," an electronic spell checker will not catch your mistake. Scanners may be used to scan line art for direct placement into page-layout programs or to create templates for use in drawing programs. Scanners also may be used for scanning photographs, but further digital image manipulation is sometimes necessary to create acceptable images for printing. Black-and-white and color laser printers can produce pages at 300-400 dots per inch currently and will certainly improve in resolution in the future. Other electronic prepress systems such as the Linotronic can put out paper or negative pages from disk at 1250 dots per inch or more. In the future, prepress systems will be linked electronically directly to presses, transferring electronic publication files directly to press plates for printing.

Beyond the printed word, publications are now already appearing as electronic files shared on networks, or on compact disks that can store giant amounts of pictorial, sound, and written information. Many of these electronic forms are interactive, allowing for greater use as educational tools at all levels.

Traditional print may lose ground to new forms. Electronic methods will continue to enhance printing as we know it, and may even come to dominate; they cannot replace it. These developments not only change the technology of publication, they force a redefinition of its meaning. Is an electronic publication — one that never appears on paper — a publication in the same sense as an article in a traditional journal? Is a computer-generated map that is plotted and produced in only a few copies equivalent as a publication to one that is printed by the thousands? As computer technology opens new avenues of reproduction, it may also increase the importance of the peer-review process in determining what is and what is not a true publication.

Reference Shelf

Any editorial office should have a shelf of reference works on subjects ranging from writing through printing — those in addition to books in the field of the publication's subject matter. You might think of this section of *Geowriting* as a shopping list for such a reference shelf; in it we list important works in various categories and compare their merits.

STYLE BOOKS

One of the main tools of any editor or writer is a style book. Yet style books are notoriously hard to use, and usually each one is intended for a single purpose and so is unlikely to meet the needs of others. Almost all editors adopt a style and either adapt it to their needs or try (wrongly) to force the editorial matter to fit the style book's mold.

Chicago Manual of Style (14th ed., 1993; 921 p.). This, too, is used for an entire stable of publications, including the *Journal of Geology*.

Manual for Writers of Term Papers, Theses, and Dissertations (5th ed., 1987; University of Chicago Press; 300 p.) by Kate L. Turabian may be the most widely used standard in college departments of geology.

Scientific Style and Format: The CBE Manual for Authors, Editors, and Publishers (6th ed., by the Style Manual Committee, Council of Biology Editors, 1994; Cambridge University Press, Cambridge, UK; 825 p.) is not merely a style book but also includes sections on special scientific conventions, journals and books, and the publishing process. It is a good starting point for, say, paleontologists, as well as biologists.

Suggestions to Authors of the Reports of the United States Geological Survey (7th ed., revised and edited by Wallace R. Hansen, 1991; Government Printing Office, Washington, D.C.; 289 p.). In geology, this is the best-known style reference. Much of the discussion is aimed at authors of U.S. Geological Survey publications, but the sections on style, stratigraphic nomenclature, and preparation of maps and figures are applicable and useful for almost any author working in the earth sciences.

U.S. Government Printing Office Style Manual (1984; Washington, D.C.; 479 p.). This is a widely used, highly useful work, with suggestions to authors, and guides to capitalization, spelling, punctuation, and so on, plus guides to the typography of many foreign languages. However, its usefulness is diluted by instructions applicable only to the Congressional Record and other specific government publications, and by the fact that it tries to be all things to all editors. You are likely to find it a handy reference but not a bible.

Water Resources Division Publications Guide — Volume I; Publications Policy and Text Preparation (1982; U.S. Geological Survey, Water Resources Division, 497 p.), by Anne J. Finch and David Aronson.

Words into Type (3rd ed., based on studies by Marjorie E. Skillin, Robert M. Gay and others, 1974; Prentice-Hall, Englewood Cliffs, N.J.; 585 p.) concerns not only style but also grammar, usage, and printing practice.

JOURNAL STYLE

Many technical journals publish their style rules every year or so in the journals themselves, as does *Science* magazine. Also, some newspaper style books are particularly useful as a starting point for making up a guide to fit your publication.

The Associated Press Stylebook and Libel Manual (Norm Goldstein, ed., 1993; Associated Press, New York; 341 p.) contains alphabetical entries for both style and word usage.

A Handbook for Scholars (by Mary-Claire van Leunen, 1978; Knopf, New York; 354 p.) is a comprehensive guide to "the mechanics of scholarly writing"; despite the pedantic title, it is a witty and authoritative source of information on how to handle citations, references, footnotes, bibliographies, format, text preparation, and the like.

The MLA Style Sheet (2nd ed., 1970; Modern Language Association of America, New York; 48 p.) is widely used by nontechnical scholarly journals, with more than 2,600,000 copies in print.

The New York Times Manual of Style and Usage: A Deskbook of Guidelines for Writers and Editors (edited by Lewis Jordan, 1976; Quadrangle/ New York Times Book Co.; 231 p.). Many of its entries concern language usage rather than house style. The *Times'* guide is alphabetical; that is, if you want to determine when and how to abbreviate "Missouri" you look up not a section on state names or on abbreviations but the word "Missouri."

TECHNICAL WRITING

Even though the principles of good writing are largely independent of subject matter, a host of books give instructions for writing about technical subjects or for technical journals.

Effective Writing for Engineers, Managers, and Scientists (2nd ed., by H.J. Tichy with Sylvia Fourdrinier, 1988; John Wiley & Sons, New York; 580 p.). If you must settle for a single book on the subject, this is it. One chapter alone will convince almost anyone: "Two dozen ways to begin."

Elements of the Scientific Paper (1985; Yale University Press, New Haven, Conn.; 130 p.), by Michael J. Katz.

Engineered Report Writing: A Manual for Scientific, Technical, and Business Writers (2nd ed., by Melba W. Murray and Hugh Hay-Roe, 1986; Pennwell Publishing Co., Tulsa, Okla.; 292 p.) Pages 141-203 are especially useful.

Handbook for Academic Authors (by Beth Luey, 1987; Cambridge University Press, Cambridge; 226 p.). A guide to writing journal articles, revising dissertations, and finding and working with a book publisher.

How to Write and Present Technical Information (2nd ed.; by Charles S. Sides, 1991; Oryx Press, Phoenix, Ariz; 182 p.)

How to Write and Publish a Scientific Paper (3rd ed., by Robert A. Day, 1988; Oryx Press, Phoenix, Ariz.; 224 p.) is the best bargain in its area and belongs near the top of your list of must-have books. Almost all science writers and editors will learn from it; all will enjoy reading it. Pages 1-41 are especially useful.

Scientific Writing for Graduate Students (Council of Biology Editors; 1989; 187 p.). Describes not only the writing process, but also preparation of figures, oral presentations, and writing dissertations and proposals.

The Scientist's Handbook for Writing Papers and Dissertations (by Antoinette M. Wilkinson, 1991; Prentice Hall, Englewood Cliffs, New Jersey; 542 p.) Covers all aspects of science writing from research to the publication process; special emphasis is placed on writing abstracts, preparing tables and graphics, and equations.

Scientists Must Write: A Guide to Better Writing for Scientists, Engineers, and Students (by Robert Barrass, 1978; Chapman & Hall, London; 176 p.). As the author says, "It is about all the ways in which writing is important to students and working scientists and engineers in helping them to remember to observe, to think, to plan, to organize, and to communicate."

Technical Communication (by Rebecca Burnett Carosso, 1986; New York, Wadsworth Publishing Co.; 622 p.) is an exhaustive discussion of report and manual writing, design techniques, and virtually anything related to technical writing.

Writing in Earth Science (by Robert L. Bates, 1988; American Geological Institute, Alexandria, Virginia; 50 p.). A how-to guide, intended for "those who are sure of their science but are not so sure of their ability to communicate it."

Writing in Nonacademic Settings (by Lee Odell and Dixie Goswami, eds., 1985; Guilford Press, New York; 553 p.) A collection of articles about writing in a variety of settings, mostly business.

Writing Successfully in Science (by Maeve O'Connor, 1991; Harper Collins Academic, London; 229 p.). Gives all the information that a writer is likely to need in preparing a research paper for publication in a scientific journal.

Several journals also deal with research and issues related to technical communication. They include *The Journal of Business and Technical Communication, The Journal of Technical Writing and Communication, Technical Communication* (a publication of the Society for Technical Communication), and *Technical Communication Quarterly*.

Associations of interest for the technical communicator include the Council on Programs in Technical and Scientific Communication, the Society for Technical Communication, the Council of Biology Editors, the Association of Earth Science Editors, and the Society for Scholarly Publishing.

EDITING

Chicago Guide to Preparing Electronic Manuscripts (1987; University of Chicago Press, Chicago; 143 p.). Describes using the word processor to edit and design for computerized typesetting.

Copyediting: A Practical Guide (by Karen Judd, 1982; William Kaufmann Inc., Los Altos, Calif.; 287 p.). Copyediting being an integral part of writing, you must learn to edit your own copy critically. This is an essential book, full of information and good sense.

Editors on Editing (3rd ed.; edited by Gerald Gross, 1993; Grove Press, New York; 377 p.). Even editors sometimes wonder how other editors work. Answers may be found in this book of 25 chapters by 25 editors, discussing such topics as copyediting, textbook editing, and dealing with authors, as well as the famous "Theory and practice of editing New Yorker articles" by Wolcott Gibbs. This book reveals many points of view about a single craft.

The Elements of Editing: A Modern Guide for Editors and Journalists (by Arthur Plotnik, 1982; Macmillan, New York; 156 p.). The author says, correctly, that "most on-the-job editorial training focuses on the employer's unique requirements and not on general editorial basics" and that "most textbooks, for all their mass, do not address the range of practical problems a new editor will face in the course of a year."

How to Copyedit Scientific Books and Journals (by Maeve O'Connor, 1986; ISI Press, Philadelphia; 150 p.).

The Scientist as Editor: Guidelines for Editors of Books and Journals (by Maeve O'Connor, 1979; John Wiley & Sons; Tunbridge Wells; 218 p.). This compact work covers a surprising range of subjects — dealing with authors, referees, and printers; launching a new journal, etc., answering in the affirmative its opening question: "Can scientists who become editors stay sane?"

Newsletters covering subjects on editing include *The Editorial Eye* and *Copy Editor*.

PUBLISHING

Book Marketing Handbook; Tips and Techniques (by Nat G. Bodian, 1980; Bowker, New York; 481 p.) If you have any interest in "the sale and promotion of scientific, technical, professional, and scholarly books and journals," you must consult this work.

Economics of Scientific Journals (ed. by D.H. Michael Bowen and others, 1982; Council of Biology Editors, Bethesda, Maryland; 106 p.) is a useful introduction to a variety of topics related to journals, including budgeting, advertising, and marketing.

Financial Management of Scientific Journals (1989; Council of Biology Editors; 112 p.). Includes chapters on distribution, advertising, working with printers, and other topics related to financial management of journals.

Folio: The Magazine for Magazine Management (Folio Magazine Publishing Corporation, New Canaan, Conn.). This monthly journal seems to be the only one of its kind; it deals not only with management but also with editing processes, circulation procedures, production methods, postal regulations, copyright, and on and on.

The Huenefeld Guide to Book Publishing (rev. 4th ed., by John Huenefeld, 1990; Mills & Sanderson, Publishers; 303 p.). Covers all facets of book publishing and starting a new venture, from editorial development, marketing, and management, to actual production.

Scientific and Technical Journals (by Jill Lambert, 1985; Clive Bingley, London; 191 p.).

Two professional societies also publish newsletters that are useful for earth-science publishing. They are the Association of Earth Science Editors and the Society for Scholarly Publishing.

USAGE

To many editors (and writers, and even those who only read), usage is the most fascinating of subjects. Almost everyone has notions on the subject, and many follow the practice of making up their own lists of usages to follow or to guard against. As a result there are a great many book-length treatments of the subject and several that are outstanding.

A Dictionary of Modern English Usage, known as "Fowler" or "M.E.U." (H.W. Fowler; revised by Sir Ernest Gowers, 1965; Oxford University

Press, Oxford; 725 p.). This classic work on English usage is the most widely used reference in editorial offices in the English-speaking world. Few editors dare neglect it, and fewer still will do so after reading entries on (say) "humour" or "split infinitive."

The Careful Writer: A Modern Guide to English Usage (by Theodore M. Bernstein, 1977; Atheneum, New York; 487 p.) is an alphabetically arranged discussion of usages. Its origins may be traced to Winners and Sinners, "a bulletin of second-guessing" that the author produced for many years from a corner of the *New York Times* news room, with the *Times* itself under review. This work is less exhaustive than Fowler but more entertaining, and it focuses on everyday problems in current American writing.

English Language and Usage in Geology: A Personal Compilation (by Dorothy H. Rayner, 1982; Leeds Geological Association, Leeds, England, 30 p.). A good guide to grammar, syntax, and idiom, plus ten brief selections of expository prose as examples.

The Handbook of Nonsexist Writing (2nd ed., by Casey Miller and Kate Swift, 1988; Harper and Row, New York; 180 p.).

Miss Thistlebottom's Hobgoblins (Theodore M. Bernstein, 1971; Farrar, Straus & Giroux, New York; 260 p.) is subtitled *The careful writer's guide to the taboos, bugbears and outmoded rules of English usage*. It exposes such superstitions as the rule against splitting infinitives, and also includes several short works on usage such as Ambrose Bierce's "Write it Right."

Plain English Handbook: A Complete Guide to Good English (6th ed. by J. Martyn Walsh and Anna Kathleen Walsh, 1972; McCormick-Mathers, Cincinnati, 216 p.). Brief, well-organized, especially useful for beginners.

Words on Words: A Dictionary for Writers and Others Who Care About Words (by John B. Bremner, 1980; Columbia University Press, New York; 406 p.) defines and discusses the usage of a number of common or particularly problematic words. Bremner, a University of Kansas journalism professor, obviously cared deeply about words and correct usage. This book reflects his concern and humor.

LITERARY STYLE

The Art of Readable Writing (rev. ed. by Rudolf Flesch, 1974; Harper, New York; 271 p.). Writers and editors who concern themselves with readability levels should at least know about Flesch's attempts to measure readability and human-interest levels. Even if you don't want to go to the trouble to make the tests (although many word-processing programs will run readability tests on your writing, making the task much easier), you will find much good advice on how to write for the reader.

The Complete Plain Words (by Sir Ernest Gowers, revised by Sidney Greenbaum and Janet Whitcut, 1988; D.R. Godine, Boston; 288 p.). Careful writers tend to despise what Americans call bureaucratese or gobbledygook, and anyone in the group will be interested in this book. It was written at the invitation of the British Treasury, and is far livelier than most commissioned works. This guide concerns "the choice and arrangement of words in such a way as to get an idea as exactly as possible out of one mind into another."

The Elements of Style (3rd ed., by William Strunk, Jr. and E.B. White, 1979; Macmillan, New York; 85 p.). The sections on grammar, style, and usage may seem overly conservative, but White's chapter on writing is among the most readable in the language. Also known as "Strunk and White," this book is a revival, by White, of a textbook in the English department at Cornell dating back to 1919 or earlier. Chapter 5, "An Approach to Style," is a classic. Everyone should own a copy of Strunk and White.

Roget's International Thesaurus (4th ed., P. M. Roget, revised by Robert L. Chapman, 1979; T. Y. Crowell Co., New York; 1,317 p.) stands in one form or another on almost everyone's reference shelf. In fact, it is so widely known that a warning is in order: despite widespread belief, this is not a dictionary of synonyms and must be used in conjunction with a good dictionary. The proper and intended use of *Roget* is to guide users from a concept (when they cannot think of the word) to the word itself. Also, many word-processing programs come with a thesaurus. Use it.

A Treasury for Word Lovers; with a Foreword by Edwin Newman (1983; ISI Press, Philadelphia, PA), by Morton S. Freeman.

Writing with Precision; How to Write so that you Cannot Possibly be Misunderstood (rev. 3rd ed., by Jefferson D. Bates, 1985; Acropolis Books, Washington, D.C.; 226 p.).

WRITING GUIDES

Books that profess to make you a better writer are a dime a dozen. Good books that will actually help your writing, particularly those that focus on science and technical topics, are much more rare.

On Writing Well: An Informal Guide to Writing Nonfiction (3rd ed., William Zinsser, 1985; Harper and Row, New York, 246 p.). Includes a particularly good chapter on technical writing. Zinsser is also the author of *Writing to Learn*, which discusses the use of writing in college curricula.

Starting from Scratch: A Different Kind of Writers' Manual (by Rita Mae Brown, 1988; Bantam Books, New York; 254 p.) is by a well-known novelist; it includes an excellent reading list.

Style: Ten Lessons in Clarity and Grace (2nd ed. by Joseph M. Williams, 1985; Scott, Foresman and Co., Glenview, Illinois; 251 p.).

Style: Toward Clarity and Grace (Joseph M. Williams, 1990; University of Chicago Press, Chicago, 208 p.).

MAPS AND PHOTOS

Cartographic Design and Production (J. S. Keates, 1973; Wiley, New York; 240 p.) discusses the graphical and technical bases of cartography, as well as map production; it is clear and complete, and includes all that you're likely to need to know in this field.

Several style manuals, such as that of the Geological Survey of Canada (1980), carry information on the preparation of illustrations.

Handbook for Scientific Photography (Alfred A. Blaker, 1977; Freeman; 319 p.) Photography poses special problems: most books on the subject seem either too general or too specialized. However, this book is neither insulting nor intimidating.

SLIDES

How to Keep an Audience Attentive, Alert, and Around for the Conclusions at a Scientific Meeting (by H. E. Clifton in *Journal of Sedimentary Petrology*, v. 48, p. 1-5; reprinted in *GSA News and Information*, 1985, v. 7, p. 88-90).

Figuratively Speaking: Techniques for Preparing and Presenting a Slide Talk (edited by Duncan Heron, 1986; American Association of Petroleum Geologists, Tulsa; 110 p.). Geologists seem particularly enamored of slide presentations. This book explains the right way to use slides.

Make the Last Slide First (by E. A. Shinn, 1981, in the *Journal of Sedimentary Petrology*, v. 51, p. 1-6).

DESIGN AND ILLUSTRATION

Bookmaking: The Illustrated Guide to Design/Production/Editing (2nd ed., by Marshall Lee.; R. Bowker Co.; 1980; 485 p.). An exhaustive reference to all aspects of book design, typography, and manufacturing for production managers.

The Bookman's Glossary (6th ed., edited by Jean Peters, 1983; Bowker, New York; 223 p.) is also helpful in book publishing.

Corporate Design Programs (by Olle Eksell, 1967; Reinhold/Studio Vista, New York; 96 p.). An editor needs to maintain consistency of design throughout a single publication, several publications, or even a larger scheme of things. Principles, problems, and solutions are discussed in terms of coordinating graphic design throughout all aspects of an organization's image: trademark, logotype, typography, letterhead, packaging, and the like.

Designer's Guide to Print Production (edited by Nancy Aldrich-Ruenzel; 1990, Watson-Guptill Publications, New York; 159 p.) by the editorial director of *Step-by-Step Graphics*, this book is just that: a step-by-step guide from preparing copy for typesetting to preparing art and photos for reproduction to pointers and printing and paper.

Editing by Design: A Guide to Effective Word-and-Picture Communication for Editors and Designers (by Jan V. White, 1982; R. Bowker Co.; 248 p.) An excellent and complete guide for the editor who must work on more than the text in preparing a book; good information on what works in book design.

Envisioning Information (by Edward R. Tufte, 1990; Graphics Press, Cheshire, Connecticut, 126 p.) discusses techniques of displaying information and includes numerous good examples. One reviewer called it "a visual Strunk and White."

The Graphics of Communication. (5th ed., by Russell N. Baird, Arthur S. Turnbull, and Duncan McDonald, 1987; Holt, Rinehart, Winston, 416 p.). An excellent source of visual communication.

Illustrating Science: Standards for Publication (Council of Biology Editors, 1988; 308 p.). Focuses particularly on preparing figures, photographs, and text for publication.

Ink on Paper 2: A Handbook of the Graphic Arts (by Edmund Arnold, 1974; Harper & Row, New York; 374 p.) tells little about ink or paper, oddly enough, but a great deal about how to get the former on the latter.

Papers for Printing: How to Choose the Right Paper at the Right Price for any Printing Job (by Mark Beach and Ken Russon, 1989; Coast to Coast Books, Inc., Portland, OR; 59 p. plus samples) provides a lot of information on selecting and specifying paper stock for all kinds of printing; useful appendices and glossary of information making you better able to communicate with your printer.

Pocket Pal: A Graphic Arts Digest for Printers and Advertising Managers (12th ed., 1979; International Paper Co., New York; 204 p.). The title of this useful book describes not only the content but the size: you really can put it in your pocket.

Steps Toward Better Scientific Illustrations (2nd ed., by Arly Allen, 1977; Allen Press, Lawrence, Kansas; 36 p.) describes the printing process primarily as it relates to the use of illustrations in scientific publications. Basic, practical information.

Typographics: A Designer's Handbook of Printing Techniques (by Michael Hutchins, 1969; Studio Vista/Reinhold, New York; 96 p.) is a concise but wide-ranging introduction to design and production. It covers production from communicating with the printer through various problems of binding (and their solutions).

The Visual Display of Quantitative Information (Edward R. Tufte, 1983; Graphics Press, Cheshire, Connecticut; 197 p.) is an influential look at the theory and practice in design of statistical graphics, charts, maps, and tables. It reproduces 75 examples of good graphical work. An excellent resource.

The periodical *Communication Arts* includes high-quality reproductions of successful design in a variety of media. Its annual collection of award-winning design is particularly helpful.

COPYRIGHT LAW

In theory, editors of scientific works should have fewer legal problems than editors of, say, crusading newspapers. However, copyright alone encompasses issues enough to justify a book on the reference shelf.

Copyright Handbook (2nd ed., by Donald F. Johnston, 1982; Bowker, New York; 381 p.) Because the United States adopted a new copyright act, this work on the subject is important. Also useful is *The Copyright Handbook – A Practical Guide* (3rd ed., by William S. Strong, 1990; The MIT Press, 249 p.)

DICTIONARIES

The essential reference for any writer or editor is a dictionary. A good one can even substitute (up to a point) as a general style manual by serving as a standard for spelling and hyphenation, tabulating proofreaders' marks, discussing principles of usage and punctuation, and defining at least the basic printing terms.

Fortunately there are a wide range of dictionaries. It is too easy to judge them solely by the copyright date on the assumption that the most recent will have the latest definitions and the newest words. Other factors may be quite as important: the presence or lack of illustrations, place names, systems of measurement, and typographic conventions. Also the philosophy of the editors and their basis for choosing terms and defining them is important; the introduction to a dictionary usually presents a forbidding appearance but reading it will be well worth your while.

Many of the following dictionaries are also available in electronic versions that you can install on your computer and use while you're writing.

In many ways the leading English dictionary is *The Oxford English Dictionary*, or *O.E.D.* (prepared by J. A. Simpson and E. S. C. Weiner, 1989; Oxford University Press; 20 volumes). The *O.E.D.* is intended to give each meaning of every word that was ever in English, and to quote the earliest known use of each. The whole is available in a small-print edition (2 volumes, 4,134 p.); a series of supplements brings the work up to date. Also there is a long series of Oxford dictionaries (*The Oxford Universal Dictionary on Historical Principles*, *The Shorter Oxford Dictionary*, *The Concise Oxford Dictionary*, and so on) working down from the 13 volumes plus supplements to one of pocket size. It is also available on CD-ROM.

Webster's Third New International Dictionary of the English Language, unabridged, is the leading "unabridged" dictionary in the United States (edited by Philip B. Gove, 1986; G.& C. Merriam Co., Springfield, Mass.; 2,662 p.) It is often compared with *Webster 2*, or the second edition of *Webster's New International Dictionary of the English Language* (edited by W. A. Neilson and others, 1957; 3,194 p.); *Webster 2* remains useful largely because the second edition specifies "proper" usage and the third is content to report usage. But *Webster 2* is out of date, out of print, and generally unavailable. Both the *O.E.D.* and *Webster* provide definitions for a given term in chronological order; that is, the last definition is the most recent and therefore often the most preferred.

DESK DICTIONARIES

The American Heritage Dictionary of the English Language (3rd ed., 1992; Houghton Mifflin Co., Boston; 2,140 p.) includes helpful discussions of controversial usages (see its "hopefully"). Well illustrated.

The Concise Oxford Dictionary (8th ed.; edited by R. E. Allen, 1990; Oxford University Press; 1,454 p.) is helpful for dealing with British writing.

Funk & Wagnalls New International Dictionary of the English Language (1987; New York, Literary Guild World Publishers; 2 vols.) stresses its usage notes.

The Random House Dictionary of the English Language–The Unabridged Edition (2nd ed., edited by Stuart Berg Glexner, 1987; Random House, New York; 2,478 p.) and *The Random House College Dictionary* (rev. ed., 1975, 1568 p.) specifies whether a word or usage is "standard."

Webster's New World Dictionary of the American Language (3rd college ed., edited by Victoria Neufeldet and David B. Guralnik; 1988; Prentice Hall, New York; 1,574 p.) stresses new science terms and etymologies—valuable points to consider. Webster's "New World" is exceptionally influential, and has been adopted as the standard desk dictionary by major newspapers and wire services.

Webster's Ninth New Collegiate Dictionary (1990; Merriam; 1,564 p.) is the first Collegiate to include usage notes, and for each word gives the date of its first known use. Thus it is more up to date than its ostensible parent, and, in addition, uses different (and generally thought superior) rules for hyphenation.

GLOSSARIES

The Concise Oxford Dictionary of Earth Sciences (by Alisa Allaby and Michael Allaby, 1990; Oxford University Press, Oxford and New York; 410 p.). Compact, authoritative, and primarily British, this work includes a 250-title bibliography as well as biographical notes on some of the great people in the earth sciences.

Dictionary of Geological Terms (3rd ed., edited by Robert L. Bates and Julie A. Jackson, 1984; Anchor Press/Doubleday, Garden City, N.Y.; 571 p.) contains about 9,000 terms and provides the nonspecialist with accurate definitions of the working vocabulary of the earth sciences.

Dictionary of Geology (6th ed., by John Challinor, edited by Antony Wyatt, 1986; University of Wales Press; 574 p.) is especially valuable in working with British geological reports (old or new). It discusses about 1,500 terms, many of them unfamiliar in the United States.

Glossary of Geology (3rd ed., edited by Robert L. Bates and Julia A. Jackson, 1987; American Geological Institute, Alexandria, Va.; 788 p.) covers 37,000 terms in geology and related earth sciences. Definitions are accompanied by guides to pronunciation and in many instances by information on first use. More than 2,000 references to the literature are included.

The International Dictionary of Geophysics (edited by S. K. Runcorn, 1967; Pergamon Press, Oxford; 2 volumes, 1,728 p.), and *Glossary of Oceanographic Terms* (edited by B. B. Baker, Jr., and others, 1966; Special Publication 35, U.S. Naval Oceanographic Office, Washington, D.C.; 204 p.) are two other specialized dictionaries.

The Oxford Dictionary for Scientific Writers and Editors (edited by Alan Isaacs, John Daintith, and Elizabeth Martin, rep. 1992, Oxford University Press, New York; 389 p.). Includes over 9,500 entries covering the physical and life sciences and reflecting current practice in scientific literature.

SCIENTIFIC NAMES

New mineral names are approved by the New Minerals & Names Committee of the International Mineralogical Association. The committee also passes on the validity of mineral species. Published summaries of their findings are included in the journals *American Mineralogist* and the *Mineralogical Magazine*.

Code of stratigraphic nomenclature, published in the May 1983 issue of the *Bulletin* of the American Association of Petroleum Geologists, provides rules for naming a rock unit. These rules are strict, and geologists should comply with them. There is no "rule board" sitting to pass on the use of formation names. Each geologist (unless working for the U.S. Geological Survey or publishing with another stringent organization) must follow the stratigraphic rules if order is to prevail. Editors must make sure names are used properly, and that new formation names follow the guidelines laid down by the Commission on Stratigraphic Names.

Glossary of Mineral Species (6th ed., by Michael Fleischer and J. A. Mandarino, 1995; Mineralogical Record Inc.; Tucson, Az.; 388 p.) is an alphabetical list of mineral species, with chemical formulas. *The Mineralogical Record* magazine carries lists of new minerals and discredited names.

An Index of Mineral Species and Varieties Arranged Chemically (2nd ed., by M. H. Hey, 1955; British Museum [Natural History], Department of Mineralogy, London; 728 p.) and its appendix (1963; 135 p.).

International Stratigraphic Guide: a Guide to Stratigraphic Classification, Terminology, and Procedure (2nd ed., edited by Amos Salvador, 1994; co-published by the International Union of Geological Sciences and the Geological Society of America; 214 p.).

Lexicon of Geologic Names of the United States (edited by M. Grace Wilmarth, 1938 (republished in 1968 by Scholarly Press); Bulletin 896, U.S. Geological Survey, Washington, D.C.; 2 volumes; 2,396 p.) is an indispensable reference in American stratigraphy. The *Lexicon* consists of rock-unit names recognized by the Survey, together with age, type section, source of original description, and similar data. (Although not intended as such, it is a useful reference for spellings of geographic names).

Lexicon of Geologic Names of the United States for 1936-1960 (edited by Grace C. Keroher, 1966; Bulletin 1200, U.S. Geological Survey; 3 volumes; 4,341 p.) and later volumes are supplemental to but do not replace "Wilmarth."

GEOGRAPHY

Place names of most large landscape features are to be found on the topographic maps of the U.S. Geological Survey.

Decisions on Geographic Names in the United States, published quarterly by the United States Board on Geographic Names, is the source of information about the correctness of a name. In it, approved new names, changes in names, and vacated names are given alphabetically by state. Information about the Board's work, or about geographic names, may be obtained from the U.S. Geological Survey (Reston, VA, 22092), whose staff assists the Board.

Webster's New Geographical Dictionary (1984; G.& C. Merriam Co., Springfield, Mass.; 1,376 p) covers geographic names.

ATLASES

The National Atlas of the United States of America (edited by Arch C. Gerlach, 1970; U.S. Geological Survey; 417 p.) is an outstanding, although expensive, atlas.

The National Geographic Atlas of the World (1990, National Geographic Society, Washington, D.C.; 300 p.) is widely used and accurate.

New International Atlas (1989; Rand McNally, Chicago; 344 p.) is also good.

The Times Atlas of the World (8th ed., 1990; Times Books, New York; 225 p.) is one of the best atlases.

SYMBOLS

Abbreviations and other symbols commonly denote units of measurement (linear, time, volumetric, and others); chemical and biochemical elements, compounds, and components; and features shown on maps. Sources of symbols and abbreviations include the International Organization for Standardization, Geneva, Switzerland, and the American National Standards Institute (1430 Broadway, New York, 10018).

"WHERE TO?"

Many "Where to find information on . . ." books and booklets have been published in geoscience. They are useful, but as the information they refer to is constantly changing, their useful life is very short. Helpful volumes include almanacs (several, such as the *World Almanac* and the *Statistical Abstract of the United States*, are published annually), encyclopedias, concordances, indexes, bibliographies, directories, thesauruses, guides to archival and unpublished material, lists of theses and dissertations, lists of publications from various publishers — federal, state, society, and private — specialized glossaries and dictionaries, special atlases, and many, many more, available generally at libraries. Consult your librarian about sticky problems, especially for areas not in your field.

ABOUT PEOPLE

American Men and Women of Science (Jaques Cattell Press/R.R. Bowker Co., New York) contains biographical detail about scientists.

Dictionary of American Biography (1980; G.C. Scribner's Sons, New York; 1,333 p.), with supplements, may also help.

The Dictionary of Scientific Biography (Charles Gillispie, ed.; 1970; Scribner, New York; 16 vols.) contains information about historic figures in science.

See the various versions of *Who's who*, such as *Who's who in engineering*, *Electrical who's who*, *Who's who in government*, *Who's who in politics*, and *Who was who* (published by Marquis Who's Who Inc., Chicago); most are published at 2-year intervals.

Memorials published by the Geological Society of America, the American Association of Petroleum Geologists, and other professional organizations are good sources of information about deceased geologists. The *New York Times* obituaries are also useful.

INDEXES

Indexing is an art that requires high skill. Purpose of the index, interests of the users, subject matter, bias of the indexer — all those enter into the making of an index. There are rules one can follow: libraries have special requirements; individual publishers often follow individual methods; society publications and technical journals frequently have in-house requirements that have been developed through the life of the publication. Study of good indexes, and practice in making your own, are probably the best ways of learning to index. Books and articles on basic techniques are available at libraries, but none exists on the specific problems involved in indexing books or other publications in earth science.

GeoRef Thesaurus (7th ed., edited by Barbara Goodman, 1994; American Geological Institute, Alexandria, Va.; 841 p.) contains more than 28,000 terms and the complete GeoRef indexing structure.

Authors and editors may find their way through the maze of making a normal index, but they are unlikely to be able to do key-word indexing for storage in a computer unless they have some reasonably specific information on what is required in making and using a computer-stored index.

The design and use of computer-system indexes involve a whole new set of concepts and techniques.

The Thesaurus of Engineering and Scientific Terms (1970, Engineers Joint Council, New York, 89 p.) is intended to be a uniform base for indexing the various branches of engineering and science. The book's subtitle describes its intent: "A list of engineering and related scientific terms and their relationships for use as a vocabulary reference in indexing and retrieving technical information." Using the thesaurus as the basic list of machine-acceptable reference words, you can construct more detailed breakdowns of specific fields, thereby creating a specialized, computer-oriented thesaurus. The International Mineralogical Association and the U.S. Office of Water Resources Research have done just that: prepared key-word thesauruses for use by the editors and indexers of all journals dealing with their subjects. Thesauruses in other fields are under way.

WRITING FOR THE MEDIA

Geomedia: A Guide for Geoscientists Who Meet the Press (by Lisa A. Rossbacher and Rex C. Buchanan, 1988; American Geological Institute, Alexandria, Virginia; 45 p.). An introduction to working with the press, written expressly for earth scientists.

Informing the People: A Public Affairs Handbook (edited by Lewis M. Helm, Ray Eldon Hiebert, Michael R. Naver, and Kenneth Rabin, 1981; Longmans, New York, 359). A collection of articles on public-information practices in government.

News Reporting: Science, Medicine, and High Technology (by Warren Burkett, 1986; Iowa State University Press, Ames; 160 p.). A primer on science writing by a long-time observer of the process.

Reflections on Science and the Media (by June Goodfield, 1981; American Association for the Advancement of Science, Washington, D.C.; 113 p.).

Science and the Mass Media (by Hillier Krieghbaum, 1967; New York University Press, New York; 242 p.). A classic text on the subject by a professor at New York University.

Scientists and Journalists: Reporting Science as News (ed. by Sharon M. Friedman, Sharon Dunwoody, and Carol Rogers, 1986; American Association for the Advancement of Science, New York; 333 p.). An excellent compendium of advice and observations by practitioners and professors of science writing. A particularly strong chapter on the audience for science, and a bibliography of research on science communication.

Selling Science: How the Press Covers Science and Technology (by Dorothy Nelkin, 1987; W.H. Freeman and Co., New York; 225 p.) is a series of essays on the relationship between science and the media by a Cornell University professor.

Writing for Story (by Jon Franklin, 1986; Mentor, New York, 284 p.). Jon Franklin has won two Pulitzer Prizes for non-fiction writing. You may disagree with his approach, but nobody writes better.

Writer's Guide to Periodicals in Earth Science

The following information is current as of February 1995.

AAPG Bulletin

ISSN 0149-1423

The *Bulletin* deals primarily with petroleum geology, but also includes articles in related energy fields. Technical journal of petroleum geology; worldwide in scope and contributions.

Edited by Kevin T. Biddle.

Editorial address: 1444 South Boulder Avenue, Tulsa, OK 74119-3604.

Telephone: 918/584-2555.

Published monthly by the American Association of Petroleum Geologists.

Total circulation: 42,000.

Regular subscription: $135.00 a year, $160.00 foreign.

American Journal of Science

ISSN 0002-9599

The *Journal* publishes papers dealing with all aspects of geology and the geosciences.

Edited by John H. Ostrom and Danny M. Rye.

Editorial address: Kline Geology Laboratory, Yale University, Box 6666, New Haven, CT 06511.

Telephone: 203/432-3131.

Published 10 times a year by the *American Journal of Science*.

Total circulation: 3,000.

Regular subscription: $60.00 a year individuals, $30.00 a year students, $125.00 a year institutions.

American Mineralogist

ISSN 0003-004X

The *American Mineralogist* publishes results of original scientific research in the general fields of mineralogy, crystallography, and petrology. Specific topics include descriptive mineralogy, mineral properties, experimental mineralogy and petrology, geochemistry, isotope mineralogy, mineralogical apparatus and techniques, mineral occurrences and deposits, paragenesis, petrography and petrogenesis, topographical mineralogy, and crystal chemistry.

Edited by Theodore Labotka and Richard J. Reeder.

Editorial address: Editors, *American Mineralogist*, 202 East Washington St., Rm. 510, Ann Arbor, MI 48104-2017.

Telephone: 313/665-2425.

Published 6 times a year by the Mineralogical Society of America.

Total circulation: 4,000.

Regular subscription: $250.00 U.S. institutions, $235.00 foreign institutions.

Annual Review of Earth and Planetary Science

ISSN 0084-6597

Edited by George W. Wetherill.

Original reviews of critical literature and current developments in earth and planetary sciences.

Editorial address: 4139 El Camino Way, Box 10139, Palo Alto, CA 94303.

Telephone: 415/493-4400.

Published by *Annual Reviews*.

$59.00 a year, $64.00 foreign.

Applied Geochemistry

ISSN 0883-2927

Applied Geochemistry is an international journal devoted to original research papers in geochemistry and cosmochemistry which have some practical application to an aspect of human endeavor, such as the search for resources, their upgrading, preservation of the environment, agriculture, and health.

Edited by Ron Fuge.

Editorial address: Ron Fuge, Institute of Earth Studies, University of Wales, Aberystwyth, Dyfed SY23 3DB, United Kingdom.

Published by Pergamon Press.

Regular subscription: $329.00 institutions.

Applied Hydrogeology

ISSN 0941-2816

Applied Hydrogeology is the official journal of the International Association of Hydrogeologists.

Edited by E. S. Simpson.

Editorial address: E. S. Simpson, Department of Hydrology and Water Resources, University of Arizona, Tucson, AZ 85721.

Published by Verlag Heinz GmbH and Co.

Regular subscription: DM 45 (IAH members), DM 90 (other individuals), DM 120 (institutions).

Boreas

ISSN 0300-9483

Boreas accepts manuscripts from all branches of Quaternary research. Aspects of the Quaternary environment, in both glaciated and non-glaciated areas, are dealt with: climate, shore displacement, glacial features, land forms, sediments, organisms and their habitat, and stratigraphical and chronological relationships.

Edited by Christian Hjort and Karstin Malmberg Persson.

Editorial address: Christian Hjort and Karston Malmberg Persson, Department of Quaternary Geology, Solvegatan 13, S-223 62 Lund, Sweden.

Telephone: 46-46-107881.

Published quarterly by Scandinavian University Press.

Regular subscription: $72.00, $128.00 institutions.

Bulletin of the Association of Engineering Geologists

ISSN 0004-5691

The *Bulletin* publishes results of original research, case histories, proceedings, and technical notes. Selected technical papers in the field of engineering geology. Topics cover site selection through project maintenance, geomorphic processes, applied and environmental geology, and hydrogeology.

Edited by Norman R. Tilford.

Editorial address: Norman R. Tilford, Department of Geology, Texas A&M University, College Station, TX 77843.

Telephone: 409/845-9682.

Published 4 times a year by the Association of Engineering Geologists.

Total circulation: 3,500.

Regular subscription: $70.00, $80.00 foreign.

Bulletin of the Seismological Society of America

ISSN 0037-1106

Publishes scientific papers on the various aspects of seismology, including investigation of specific earthquakes, theoretical and observational studies of seismic waves, and earthquake hazard and risk estimation.

Edited by Charles M. Langston.

Editorial address: Charles M. Langston, Department of Geosciences, Pennsylvania State University, University Park, PA 16802.

Published bimonthly by The Seismological Society of America.

Total circulation: 2,500.

Regular subscription: $135.00 North America, $145.00 foreign.

Canadian Journal of Earth Sciences

ISSN 0008-4077

The *Journal* publishes the results of research in various fields of the earth sciences. Originality of ideas and excellence of research and presentation are the main criteria.

Edited by D. J. W. Piper.

Editorial address: Dr. D. J. W. Piper, Editor, *Canadian Journal of Earth Sciences*, Bedford Institute of Oceanography, MS 235, P.O. Box 1006, Dartmouth, Nova Scotia, Canada B2Y 4A2.

Telephone: 902/426-2735.

Published monthly by the National Research Council of Canada.

Total circulation: 4,600.

Regular subscription: $128.00 (Canadian) individuals, $398.00 (Canadian) institutions.

The Canadian Mineralogist

ISSN 0008-4476

The *Canadian Mineralogist* covers results of original research in mineralogy, petrology, crystallography, geochemistry, ore deposits, and mineral deposits.

Edited by R. F. Martin.

Editorial address: Department of Earth & Planetary Sciences, McGill University, 3450 University Street, Montreal, Quebec, Canada H3A 2A7.

Telephone: 514/392-5835.

Published 4 times a year by the Mineralogical Association of Canada.

Total circulation: 2,000.

Regular subscription: $60.00 (Canadian) individuals, $190.00 (Canadian) institutions.

Chemical Geology

ISSN 0009-2541

Chemical Geology publishes original studies and comprehensive reviews in organic and inorganic geochemistry.

Edited by W. S. Fyfe.

Editorial address: P. Deines, Department of Geosciences, 209 Deike Building, The Pennsylvania State University, University Park, PA 16802.

Telephone: 519/679-3187 (Fyfe).

Published 28 times a year by Elsevier Science Publishers.

Regular subscription: Dfl 2,527 a year.

Clays and Clay Minerals

ISSN 0009-8604

Clays and Clay Minerals publishes the latest advances in research and technology that deal with clays and other fine-grained minerals, and their role in mineralogy, crystallography, geology, geochemistry, sedimentology, soil science and soil mechanics, physical chemistry, colloid chemistry, ceramics, and petroleum and foundry engineering.

Edited by Ray E. Ferrell.

Editorial address: Ray E. Ferrell, Department of Geology and Geophysics, Louisiana State University, Baton Rouge, LA 70803.

Published 6 times a year by the Clay Minerals Society.

Total circulation: 1,800.

Regular subscription: $140.00 institutions, $155.00 foreign.

Computers & Geosciences

ISSN 0098-3004

Computers & Geosciences serves as a public medium for the exchange of ideas between the geological and computer sciences.

Edited by D.F. Merriam.

Editorial address: D.F. Merriam, Kansas Geological Survey, University of Kansas, Lawrence, KS 66047-3726.

Telephone: 913/864-3965.

Published 10 times a year by Elsevier Science Ltd.

Regular subscription: $930.00 institutions.

Contributions to Mineralogy and Petrology

ISSN 0010-7999

Contributions publishes original articles presenting essentially new scientific findings on geochemistry. Includes isotope geology; the petrology and genesis of igneous, metamorphic, and sedimentary rocks; experimental petrology and mineralogy; and distribution and significance of elements and their isotopes in rocks.

Edited by J. Hoefs and T. L. Grove.

Editorial address: T. L. Grove, 54-1220 Department of Earth, Atmospheric and Planetary Sciences, Massachusetts Institute of Technology, Cambridge, MA 02139.

Telephone: 617-253-2878.

Published 12 times a year by Springer-Verlag.

Regular subscription: DM 2,988 ($2,043).

Earth

ISSN 1056-148X

Covers different earth sciences topics for a general audience, including geology, paleontology, oceanography, meteorology, and mineralogy, as well as travel and exploration.

Edited by Tom Yulsman.

Editorial address: Box 1612, Waukesha, WI 53187.

Tel: 414-796-8776.

Total circulation: 71,800.

Published 6 times a year by Kalmbach Publishing Co.

Regular subscription: $19.95, $26.00 foreign.

Earth and Planetary Science Letters

ISSN 0012-821X

Earth and Planetary Science Letters covers research into all aspects of lunar studies, plate tectonics, ocean floor spreading, and continental drift, as well as basic studies of the physical, chemical, and mechanical properties of the Earth's crust and mantle, the atmosphere, and the hydrosphere.

Edited by F. Albarede, U. R. Christensen, M. Kastner, C. Langmuir, P. Tapponnier, R. VanDerVoo.

Editorial address: Professor M. Kastner, Geological Research Division, Scripps Institution of Oceanography, University of California, La Jolla, CA 92093.

Telephone: 619/534-2065 (Kastner).

Published 28 times a year by Elsevier Science Publishers.

Regular subscription: Dfl 2,366 a year.

Economic Geology and The Bulletin of the Society of Economic Geologists (formerly Economic Geology)

ISSN 0361-0128

Economic Geology concerns all aspects of economic geology and all classes of mineral deposits.

Edited by Brian J. Skinner.

Editorial address: P.O. Box 208110, New Haven, CT 06520.

Telephone: 203/432-3166.

Published 8 times a year by *Economic Geology* Publishing Co.

Total circulation: 7,000.

Regular subscription: $70.00 a year, $115.00 institutions.

Engineering Geology

ISSN 0013-7952

Engineering Geology is an international journal for the publication of original studies, case histories, and comprehensive reviews in the field of engineering geology.

Edited by M. Arnould and E.L. Krinitzsky.

Editorial address: Engineering Geology, P.O. Box 1930, 1000 BX Amsterdam, The Netherlands.

FAX: 31-20-5862696.

Published by Elsevier Science Publishers.

Subscription: For rates, contact Elsevier Science B.V., Journal Department, P.O. Box 211, 1000 AE Amsterdam, The Netherlands.

Environmental Geology

ISSN 0943-0105

Environmental Geology is an international multidisciplinary journal concerned with all aspects of interaction between humans, ecosystems and the Earth. This journal is the continuation of *Environmental Geology* and *Water Science*, which merged in 1993.

Edited by Philip E. LaMoreaux, Sr.

Editoral address: 2610 University Blvd. (35401), P.O. Box 2310, Tuscaloosa, AL 35403.

Telephone: 205/752-5543.

Published 8 times a year by Springer International.

Regular subscription: $344.00 (U.S.)

Eos

ISSN 0096-3941

For geophysicists; carries articles on recent research, news, employment opportunities, meeting programs and reports, announcements of grants and fellowships, and AGU activities.

Edited by A.F. Spilhaus, Jr.

Editorial address: 2000 Florida Avenue NW, Washington, D.C. 20009.

Telephone: 202/462-6900.

Total circulation: 27,000.

Published weekly by the American Geophysical Union.

Regular subscription: $190.00, $205.00 non-members.

Episodes

ISSN 0705-3797

Episodes, a global news magazine for the geosciences, covers developments of regional and global importance in the earth sciences.

Edited by Rodney Walshaw.

Editorial address: British Geological Survey, Keyworth, Nottingham NG12566, United Kingdom.

Telephone: +44(0)602-363-100.

Published 4 times a year by the International Union of Geological Sciences.

Total circulation: 3,000.

Regular subscription: $24.00 a year.

European Journal of Mineralogy

ISSN 0935-1221

The *EJM* publishes original papers, review articles, and short notes on all mineralogical fields like mineralogy, crystallography, petrology, geochemistry, and ore deposits, including applied and technical mineralogy as well as related fields.

Edited by C. Chopin, W.V. Maresch, F.P. Sassi.

Editorial address: C. Chopin, Laboratoire de Geologie, Ecole Normale Superieure, 24, rue Lhomond, F-75005 Paris, France.

Published by E. Schweizerbart'sche Verlagsbuchhandlung (Nagele u. Obermiller).

Subscription: For rates, contact E. Schweizerbart'sche Verlagsbuchhandlung (Nagele u. Obermiller), Johannesstr. 3A, D-70176 Stuttgart, Germany.

Geochimica et Cosmochimica Acta

ISSN 0016-7037

Geochimica et Cosmochimica Acta publishes papers in the fields of geochemistry and cosmochemistry; original research covering the entire spectrum of geochemistry and cosmochemistry, incorporating chemistry, geology, physics, and astronomy.

Edited by Gunter Faure.

Editorial address: 60Y Pressey Hall, 1070 Carmack Road, The Ohio State University, Columbus, OH 43210.

Published 24 times yearly by Pergamon Press for the Geochemical Society.

Total circulation: 3,800.

Regular subscription: $895.00 a year.

Geoderma

ISSN 0016-7061

The primary purpose of *Geoderma* is to stimulate wide interdisciplinary cooperation and understanding among workers in the different fields of pedology.

Edited by E.T. Elliott, J.A. McKeague, D.L. Sparks, J. Bouma.

Editorial address: Geoderma, P.O. Box 1930, 1000 BX Amsterdam, The Netherlands.

FAX: 31-20-4852696.

Published 11 times a year by Elsevier Science Publishers.

Subscription: For rates, contact Elsevier Science B.V., Journal Department, P.O. Box 211, 1000 AE Amsterdam, The Netherlands.

Geological Magazine

ISSN 0016-7568

Geological Magazine covers the entire field of earth sciences and welcomes short papers that, provided they are of scientific merit, are controversial in nature.

Edited by C. P. Hughes, N. H. Woodcock, I. N. McCave, M. J. Bickle.

Editorial address: *Geological Magazine*, Department of Earth Sciences, Downing Street, Cambridge, England CB2 3EQ.

Published 6 times a year by Cambridge University Press.

Regular subscription: $249 a year.

Geological Society of America Bulletin

ISSN 0016-7606

The *Bulletin* publishes research articles of broad geologic interest. Any aspect of geology is welcome. Presents papers on the results of international research on all earth science disciplines.

Edited by John E. Costa and Arthur G. Sylvester.

Editorial address: 3300 Penrose Place, Box 9140, Boulder, CO 80301.

Telephone: 303/447-2020.

Published monthly by the Geological Society of America.

Total circulation: 9,500.

Regular subscription: $185.00 a year, $195.00 foreign.

Geology

ISSN 0091-7613

Geology publishes basic and applied papers in the earth sciences, emphasizing the innovative and provocative; topical scientific papers on all earth science disciplines worldwide.

Edited by David M. Fountain and Henry T. Mullins.

Editorial address: 3300 Penrose Place, Box 9140, Boulder, CO 80301.

Telephone: 303/447-2020.

Published monthly by the Geological Society of America.

Total circulation: 10,500.

Regular subscription: $150 non-members, $160.00 foreign a year.

Geology Today

ISSN 0266-6979

Geology Today publishes articles on pure and applied geology for amateur and professional geologists. Topics include tectonics, radiometric dating, mining and drilling, seismology, underwater minerals, planetary geology, fossils, and the history of geology. Includes letters, book reviews, news.

Edited by Peter J. Smith.

Editorial address: Peter J. Smith, 32 St. James Close, Hanslope, Milton Keynes, England MK19 7LF.

Published 6 times a year by Blackwell Scientific Publications.

Regular subscription: L75, L85 foreign.

Geophysical Journal International

ISSN 0956-540X

Geophysical Journal International accepts papers on all geophysical subjects; work on the upper atmosphere is included when it is associated with the Earth, as for example magnetic effects at the Earth due to upper atmospheric occurrences.

Edited by M.A. Khan.

Editorial address: G.L. Choy, United States Geological Survey, National Earthquake Information Center, Box 25046, MS 967, Denver Federal Center, Denver, CO 80225.

Published monthly for the Royal Astronomical Society, Deutsche Geophysikalische Gesellschaft, and the European Geophysical Society by Blackwell Scientific Publications.

Regular subscription: $872.00.

Geophysical Research Letters

ISSN 0094-8276

Geophysical Research Letters publishes short, timely articles of broad geophysical interest; provides a forum for the rapid dissemination of current research of broad geophysical interest.

Edited by Garrett Brass.

Editorial address: 2000 Florida Avenue NW, Washington, D.C. 20009.

Telephone: 202/462-6900.

Published monthly by American Geophysical Union.

Total circulation: 4,500.

Regular subscription: $55.00 members ($74.00 foreign), $498.00 non-members ($517.00 foreign).

Geophysics

ISSN 0016-8033

Geophysics concerns petroleum, mining, or engineering geophysics, and fundamental scientific principles basic to geophysical exploration.

Edited by Bob A. Hardage.

Editorial address: Box 702740, Tulsa, OK 74170.

Telephone: 918/493-3516.

Published monthly by the Society of Exploration Geophysicists.

Total circulation: 19,000.

Regular subscription: $250.00, $265.00 foreign.

Geoscience Canada

ISSN 0315-0941

Contains state-of-the-art reviews, summaries of recent developments, and conference reports.

Edited by Phil C. Thurston.

Editorial address: Precambrian Geoscience Section, Ontario Geological Survey, 933 Ramsey Lake Road, Sudbury, Ontario, Canada P3E 6B5.

Telephone: 705/670-5976.

Published 4 times a year by the Geological Association of Canada.

Total circulation: 3,300.

Regular subscription: $40.00 (Canadian) a year.

Geotechnique

ISSN 0016-8505

Studies work in theoretical and practical geotechnical engineering.

Published 4 times a year by Institution of Civil Engineers.

Total circulation: 3,500

Regular subscription: L 125 ($250).

Geotimes

ISSN 0016-8556

Geotimes publishes news for geoscientists of professional activities, including technical meetings, educational programs, and field and laboratory research. *Geotimes* also includes a calendar of meetings and listings of new books, maps, and software.

Edited by Julia A. Jackson.

Editorial address: 4220 King Street, Alexandria, VA 22302.

Telephone: 703/379-2480.

Published monthly by the American Geological Institute.

Total circulation: 10,000.

Regular subscription: $32.95 individual ($47.95 foreign), $24.95 members of AGI member societies ($38.95 foreign), $14.95 students ($24.95 foreign), $36.95 institutions ($51.95 foreign).

Ground Water

ISSN 0017-467X

Ground Water publishes technical papers on any aspect of hydrology or ground-water geology. Includes letters, book reviews, annual index.

Edited by John Bredehoeft.

Editorial address: 234 Scenic Drive, La Honda, CA 94020.

Telephone: 800/638-7229.

Published 6 times a year by the Ground Water Publishing Co.

Total circulation: 15,500.

Regular subscription: $90.00 a year, $110.00 foreign.

Hydrological Sciences Journal

ISSN 0262-6667

Edited by Terence O'Donnell.

Editorial address: IAHS, Institute of Hydrology, Wallingford, Oxon, OX10 8BB, United Kingdom.

Published 6 times a year by Blackwell Scientific Publications Ltd.

Circulation: 1,000.

Regular subscription: L85 ($155.00).

Industrial Minerals

ISSN 0019-8544

Industrial Minerals covers non-fuel and non-metallic minerals from an economic and marketing point of view and includes articles on geology, mineral processing, and end-uses; covers non-metallic mineral producers by country.

Edited by Joyce Griffiths.

Editorial address: 220 5th Avenue, 19 Floor, New York, NY 10001.

Telephone: 800/638-2525.

Published monthly by Metal Bulletin Inc.

Total circulation: 3,600.

Regular subscription: $348.00.

International Journal of Coal Geology

ISSN 0166-5162

The *International Journal of Coal Geology* is committed to treating the basic and applied aspects of the geology and petrology of coal in a scholarly manner.

Edited by R.R. Dutcher.

Editorial address: R.R. Dutcher, Southern Illinois University, Dept. of Geology, Carbondale, IL 62901.

Telephone: 618/536-6666.

Published twice a year by Elsevier Science Publishers.

Subscription: For rates, contact Elsevier Science B.V., Journal Department, P.O. Box 211, 1000 AE Amsterdam, The Netherlands.

International Journal of Rock Mechanics and Mining Sciences and Geomechanics Abstracts

ISSN 0148-9062

The *Journal* is concerned with original research, new developments, and case studies in rock mechanics and engineering.

Edited by J.A. Hudson.

Editorial address: J.A. Hudson, 7 The Quadrangle, Welwyn Garden City, Herts AL8 6SG, United Kingdom.

FAX: 0171-589-6806.

Published 6 times a year by Elsevier Science Ltd.

Regular subscription: $976.00 institutions. For associated personal rates, contact Elsevier Science Inc., 660 White Plains Road, Tarrytown, NY 10591-5153.

Journal of Coastal Research

ISSN 0749-0208

The *Journal* covers all aspects of coastal research: geology, geomorphology, climate, littoral oceanography, hydrography, coastal hydrologics, environmental-resource management, engineering and remote sensing. Encourages the dissemination of knowledge and understanding of the coastal zone by promoting cooperation and communication between specialists in different disciplines.

Edited by Charles W. Finkl.

Editorial address: Box 2473, Colee Station, Ft. Lauderdale, FL 33303.

Telephone: 305/523-6768.

Published 4 times a year by the Coastal Education and Research Foundation Inc.

Total circulation: 800.

Regular subscription: $58.00 individual ($68.00 foreign), $125.00 institutions ($135.00 foreign), $43.00 students ($53.00 foreign).

Journal of Foraminiferal Research

ISSN 0096-1191

The *Journal* publishes original papers dealing with foraminifera and allied groups of organisms. Review articles are encouraged as are research articles of international interest.

Edited by Paul Loubere.

Editorial address: Paul Loubere, Department of Geology, Northern Illinois University, Dekalb, IL 60115.

Telephone: 815/753-7949.

Published 4 times a year by the Cushman Foundation for Foraminiferal Research.

Total circulation: 800.

Regular subscription: $80.00 a year.

Journal of Geochemical Exploration

ISSN 0375-6742

The *Journal* covers all aspects of the application of geochemistry to the exploration and study of mineral resources, and related fields, including the geochemistry of the environment.

Edited by E.M. Cameron.

Editorial address: Eion M. Cameron, Derry Laboratory, Department of Geology, University of Ottawa, Ottawa, K1N 6N5, Ontario, Canada.

Published twice a year by Elsevier Science Publishers.

Subscription: For rates, contact Elsevier Science B.V., Journal Department, P.O. Box 211, 1000 AE Amsterdam, The Netherlands.

Journal of Geological Education

ISSN 0022-1368

The *Journal* publishes review articles and covers education in geology and the earth sciences, history of geology, and the use of computers. Fosters improvement in teaching of earth sciences at all levels of instruction, emphasizes the cultural and environmental significance of the field, and disseminates information on related topics of interest.

Edited by James H. Shea.

Editorial address: University of Wisconsin-Parkside, Box Number 2,000, Kenosha, WI 53141.

Telephone: 414/553-2236.

Published 5 times a year by the National Association of Geology Teachers.

Total circulation: 2,800.

Regular subscription: $33.00, $37.00 foreign.

Journal of Geology

ISSN 0022-1376

The *Journal* welcomes contributions dealing with any aspect of geology, especially papers of wide appeal to geologists, papers dealing with new concepts, and papers that apply new approaches or methods to the solution of geologic problems. It does not publish papers that are overly specialized, merely descriptive, or only of local geographic significance.

Edited by Alfred Anderson, Jr. and Robert C. Newton.

Editorial address: Henry Hinds Laboratory, 5734 S. Ellis Ave., The University of Chicago, Chicago, IL 60637.

Telephone: 312/962-7896.

Published 6 times a year by the University of Chicago Press.

Total circulation: 2,400.

Regular subscription: $45.00 a year, $78.00 institutions, $20 students.

Journal of Geophysical Research

ISSN 0196-6928

The *Journal* is published in three sections. Section A covers solar and interplanetary physics, aeronomy, planetary atmospheres, and exterior planetary magnetism. Section B covers the physics and chemistry of the solid Earth, planetology, geodesy, geothermics, and material science, and includes separate supplements from the Lunar and Planetary Science Conference. Section C/D covers the physics and chemistry of the atmosphere, hydrosphere, air-sea interface, oceans, and ocean basins.

Edited by William J. Hinze.

Editorial address: For current address of earth sciences editor, call or write the American Geophysical Union at 2000 Florida Ave., NW, Washington, D.C. 20009.

Telephone: 202/462-6900.

Published monthly by the American Geophysical Union.

Total circulation: 9,500.

Regular subscription: $75 to $1,680 depending on membership and section.

Journal of Hydraulic Engineering

ISSN 0733-9429

Edited by E. John List.

Editorial address: Journals Department, 345 East 47th Street, New York, NY 10017.

Published monthly by the American Society of Civil Engineers.

Circulation: 5,300.

Regular subscription: $49.00 members ($73.00 foreign), $196.00 nonmembers ($220.00 foreign).

Journal of Hydrology

ISSN 0022-1694

Presents original studies, research results, and reviews on the chemical and physical aspects of surface and ground-water hydrology, hydrometeorology, hydrogeology, parametric and stochastic hydrology, agrohydrology, hydrology of arid zones, and applied hydrology.

Edited by David R. Maidment, George H. Davis, J.S.G. McCulloch, J.E. Nash.

Editorial office: Center for Research in Water Resources, Balcones Research Center, Bldg. 119, University of Texas, Austin, TX 78712.

Tel: (512)471-3131 (David R. Maidment).

Published monthly by Elsevier Science Publishers.

Regular subscription: Dfl 3,900.

Journal of Metamorphic Geology

ISSN 0263-4929

Covers entire range of metamorphic studies from the scale of the individual crystal to that of the lithospheric plate; includes geochemistry.

Edited by M. Brown, D. Robinson, J. Selverstone, R.H. Vernon.

Editorial office: M. Brown, Department of Geology, University of Maryland, College Park, Maryland 20742.

Published bimonthly by Blackwell Scientific Publications.

Regular subscription: $100.00 individuals, $435.00 institutions.

Journal of Paleontology

ISSN 0022-3360

The *Journal* publishes contributions in all fields of paleontology, including invertebrate and vertebrate paleontology, micropaleontology, and paleobotany, emphasizing taxonomic, biostratigraphic, paleoecological, paleoclimatological, or paleobiogeographic aspects.

Edited by Don C. Steinker.

Editorial address: Managing Editor, Department of Geology, Bowling Green State University, Bowling Green, OH 43403.

Telephone: 419/372-2886.

Published 6 times a year by the Paleontological Society.

Total circulation: 3,500.

Regular subscription: $99.00 a year for non-members.

Journal of Petroleum Geology

ISSN 0141-6421

The *Journal* covers petroleum geology, including exploration and recovery techniques, field case histories, and hydrocarbon studies. Review papers are also welcome.

Edited by E. N. Tiratsoo.

Editorial address: Box 21, Beaconsfield, Bucks, England HP9 1NS.

Telephone: 0494-675139.

Published 4 times a year by Scientific Press Ltd.

Total circulation: 2,000.

Regular subscription: $264.00 a year.

Journal of Petrology

ISSN 0022-3530

The *Journal* presents papers on the physics and chemistry of rocks, experimental petrology and mineralogy, rock-forming minerals and their paragenesis, microstructure of rocks, and isotope geochemistry and geochronology as applied to problems of petrogenesis.

Edited by B. G. J. Upton.

Editorial address: B. G. J. Upton, Department of Geology and Geophysics, Grant Institute, West Mains Road, Edinburgh, Great Britain EH9 3JW.

Published 6 times a year by Oxford University Press.

Total circulation: 1,700.

Regular subscription: L160 ($290.00).

Journal of Sedimentary Research

ISSN 1073-130X: *Section A: Sedimentary Petrology and Processes*

ISSN 1073-1318: *Section B: Stratigraphy and Global Studies*

The *Journal* publishes recent advances in the study of ancient and modern sediments, including petrology and petrography of carbonate and noncarbonate rocks and sediments, sedimentation processes, characterization of depositional systems and environments, diagenesis, and sedimentary tectonics. Related topics, such as sedimentary geochemistry, sedimentary mineralogy, clay mineralogy, coal mineralogy, trace fossils, paleocurrents, and evaporite sedimentology, are also welcome. Covers the basic research and most recent advances in the study of sediments, through papers, discussions, and replies to previously published papers.

Edited by J. Southard.

Editorial address: Department of Earth, Atmospheric, and Planetary Sciences, Rm. 54-1026, Massachusetts Institute of Technology, Cambridge, MA 02139.

Telephone: 617/253-7848.

Published by (SEPM) Society for Sedimentary Geology; four issues of each section are published in a year.

Total circulation: 5,000.

Regular subscription: $154.00 non-members.

Journal of Structural Geology

ISSN 0191-8141

The *Journal* publishes original research and review articles in structural geology and tectonics. Specific topics include deformation phenomena and processes on all scales, analysis of deformation elements, structural associations, microfabrics, experimental deformation, strain analysis, seismotectonics, regional structural geology, and global tectonics. Covers all aspects and processes of deformation in rocks, including folds, fracture and fabrics, structural associations in orogenic rocks, strike-slip zones, and related phenomena.

Edited by S. H. Treagus.

Editorial address: S. H. Treagus, Department of Geology, The University, Manchester, England M13 9PL.

Telephone: 061/273-7121.

Published 12 times a year by Pergamon Press.

Total circulation: 2,000.

Regular subscription: $640.00 (U.S.)

Journal of the Geological Society

ISSN 0016-7649

The *Journal* publishes articles dealing with any aspect of geology. Occasionally an issue is devoted to a single topic.

Edited by M. J. Le Bas.

Editorial address: Ms. A. Hills, Geological Society Publishing House, Unit 7, Brassmill Enterprise Centre, Brassmill Lane, Bath, United Kingdom BA1 3JN.

Telephone: 0225-445046.

Published 6 times a year by Geological Society Publishing House.

Total circulation: 4,900.

Regular subscription: L241 ($486.00), L289 foreign.

Journal of Volcanology and Geothermal Research

ISSN 0377-0273

The *Journal* publishes papers that deal with the geophysical, geochemical, petrologic, economic, environmental, and tectonic aspects of geothermal and volcanologic research. Specific studies may concern magma genesis and evolution, ore deposits related to volcanic rocks and magmas, volcano surveillance, exploration for geothermal resources, isotope studies, volcanic gas studies, and other related topics. Provides volcanologists, petrologists, and geochemists with a source of information and an outlet for rapid publication of papers in the field.

Edited by Professor Bruce D. Marsh.

Editorial address: Bruce D. Marsh, Department of Earth and Planetary Sciences, Johns Hopkins University, Baltimore, MD 21218.

Published 20 times a year by Elsevier Science Publishers.

Regular subscription: Dfl 1,690 a year.

The Leading Edge

ISSN 1070-4856

The *Leading Edge* publishes news about the geophysical industry and the Society of Exploration Geophysicists and includes technical articles, case histories, and historical profiles of leading geophysicists.

Edited by Dean Clark.

Editorial address: Box 702740, Tulsa, OK 74170-2740.

Telephone: 918/493-3516.

Published monthly by the Society of Exploration Geophysicists.

Total circulation: 20,205.

Regular subscription: $70.00, $85.00 foreign.

Lethaia

ISSN 0024-1164

Lethaia publishes papers of international interest in paleontology and stratigraphy. It is an international journal of paleontology and stratigraphy.

Edited by Christina Franzen and Stefan Bengston.

Editorial address: Dr. Lars Ramskold, Department of Palaeozoology, Swedish Museum of Natural History, Box 50007, S-104 05, Stockholm, Sweden.

Telephone: 46-46-107870.

Published 4 times a year by Scandinavian University Press.

Total circulation: 1,300.

Regular subscription: $145.00.

Limnology and Oceanography

ISSN 0024-3590

Covers all areas of aquatic research (except fisheries and pollution) at research level or advanced graduate studies.

Managing editor: Raelyn Cole.

Editorial office: Limnology and Oceanography, School of Oceanography WB-10, University of Washington, Seattle 98195.

Tel: 206/543-8655.

Published 8 times a year by the American Society of Limnology and Oceanography.

Regular subscription: $160.00.

Lithos

ISSN 0024-4937

Lithos publishes original research papers on mineralogy, petrology, and geochemistry, emphasizing the application of mineralogy and geochemistry to petrogenetic problems.

Edited by R. Gorbatschev.

Editorial address: R. Gorbatschev, Department of Mineralogy and Petrology, Institute of Geology, University of Lund, Solvegatan 13, S-22362 Lund, Sweden.

Published 8 times a year (2 vols., 4 nos. in each vol.) by Elsevier Science Publishers.

Total circulation: 1,000.

Regular subscription: Dfl 620.

Marine Geology

ISSN 0025-3227

Marine Geology is an international medium for the publication of original studies and comprehensive reviews in the fields of marine geology, geochemistry, and geophysics.

Edited by M.A. Arthur and H. Chamley.

Editorial address: M.A. Arthur, Department of Geosciences, Pennsylvania State University, 503 Deike Building, University Park, PA 16802-2714.

Telephone: 814/865-6711.

Published monthly by Elsevier Science Publishers.

Regular subscription: $1385.00.

Marine Micropaleontology

ISSN 0377-8398

Publishes results of research in all fields of marine micropaleontology of the ocean basins and continents, including paleoceanography, evolution, ecology and paleoecology, biology and paleobiology, biochronology, paleoclimatology, taphonomy, and the systematic relationships of higher taxa.

Edited by J. Lipps, H. Thierstein.

Editorial address: J. H. Lipps, Museum of Paleontology, University of California, Berkeley, CA 94720.

Telephone: 510/642-9006.

Published 8 times a year (2 vols., 4 nos. in each vol.) by Elsevier Science Publishers.

Regular subscription: Dfl 692 a year.

Mathematical Geology

ISSN 0882-8121

Mathematical Geology publishes papers on the application and use of mathematics, statistics, probability, and computers in the earth sciences.

Edited by Robert Ehrlich.

Editorial address: Robert Ehrlich, Department of Geological Sciences, University of South Carolina, Columbia, SC 29208.

Published 8 times a year by Plenum Publishing Corp.

Total circulation: 1,000.

Regular subscription: $425.00, $495.00 foreign.

Meteoritics

ISSN 0026-1114

Meteoritics publishes research papers dealing with lunar samples, interplanetary dust, tektites, impacts and asteroids, meteorites, craters, planetology, cosmochemistry, and meteor astronomy.

Edited by Derek W. Sears.

Editorial address: Derek W. Sears, Department of Chemistry and Biochemistry, University of Arkansas, Fayetteville, AR 72701.

Published 5 times a year by the Meteoritical Society, University of Arkansas.

Total circulation: 1,000.

Regular subscription: $150.00 institutions.

Micropaleontology

ISSN 0026-2803

Micropaleontology concerns all groups of microfossils. Contains international research on stratigraphy, systematics, morphology, paleobiology, and paleoecology of all microorganisms with fossilized hard parts.

Edited by John A. van Couvering.

Editorial address: Central Park West at 79th St., New York, NY 10024.

Telephone: 212/769-5656.

Published 4 times a year by the American Museum of Natural History.

Total circulation: 1,000.

Regular subscription: $70.00 individuals, $140.00 institutions.

Mineralogical Magazine

ISSN 0026-461X

Concerns mineralogy, petrology, geochemistry, and extraterrestrial material.

Edited by A. M. Clark.

Editorial address: 41 Queen's Gate, London, England SW7 5HR.

Telephone: 01/584-7516.

Published 4 times a year by the Mineralogical Society.

Total circulation: 2,000.

Regular subscription: L125 ($215.00 U.S.).

Mining Engineering

ISSN 0026-5187

Mining Engineering is directed to engineering professionals in the mining industry. Articles and news about finding, mining, and processing metallic and nonmetallic minerals and coal are welcome.

Edited by R. L. White.

Editorial address: Society of Mining Engineers, Caller D, Littleton, CO 80127.

Telephone: 303/973-9550.

Published monthly by the Society for Mining, Metallurgy, and Exploration.

Total circulation: 22,000.

Regular subscription: $80.00 a year.

Modern Geology

ISSN 0026-7775

Modern Geology reports experimental and theoretical findings and has original papers, reviews and short communications on developments in the earth and planetary sciences. It aims to preserve a balanced perspective of the physical, chemical and biological spheres and to identify major trends and areas of growth in the field.

Edited by Douglas Palmer and Hugh S. Torrens.

Editorial address: Hugh S. Torrens, Department of Geology, University of Keele, Staffordshire, ST5 5BG, United Kingdom.

Published 4 times a year by Gordon & Breach Science Publishers.

Regular subscription: $281.00 a year.

Nature

ISSN 0028-0836

Nature publishes original scientific research reports, review articles surveying recent developments in specific disciplines, short contributions, letters, and commentary. Provides comprehensive, in-depth coverage of scientific news and issues throughout the world, including relevant medical and legislative issues.

Edited by John Maddox.

Editorial address: 4 Little Essex Street, London, England WC2R 3LF.

Telephone: 01/836-6633.

Published weekly by Macmillan Magazines Ltd.

Total circulation: 30,821.

Regular subscription: $145.00 (U.S.), individual, $425.00 (U.S.) institutional/corporate a year.

Oil & Gas Journal

ISSN 0030-1388

Covers international petroleum news and technology.

Edited by John L. Kennedy

Editorial address: P.O. Box 1941, Houston, TX 77251.

Published weekly by Pennwell Publishing Co.

Regular subscription: $69.00, $127.00 non-industry.

Organic Geochemistry

ISSN 0146-6380

Organic Geochemistry was established in response to the need for a specialized medium for the rapid publication of research in this highly interdisciplinary field. Contributions covering a wide spectrum of subjects in the geosciences broadly based on organic chemistry (including molecular and isotopic geochemistry), and involving geology, biogeochemistry, environmental geochemistry, chemical oceanography, and hydrology are welcome.

Edited by J.A. Curale and A.G. Douglas.

Editorial address: J.A. Curale, Unocal Energy Resources Division, Unocal Corp., P.O. Box 76, Brea, CA 92621.

Published by Pergamon Press.

Regular subscription: $1185.00 institutions.

Palaeogeography, Palaeoclimatology, Palaeoecology

ISSN 0031-0182

Edited by P. DeDeckker, C. Newton, and F. Surlyk.

Editorial address: *Palaeogeography, Palaeoclimatology, Palaeoecology*, P.O. Box 1930, 1000 BX Amsterdam, The Netherlands.

Telephone: 61-6-2492056 (P. DeDeckker).

Published monthly by Elsevier Science Publishers.

Regular subscription: $1236.00 a year.

Palaeontology

ISSN 0031-0239

Palaeontology publishes papers on all aspects of paleontology, including paleobotany, stratigraphy, and paleobiology. Abstracts in English.

Edited by Dr. R. M. Owens.

Editorial address: R. M. Owens, Department of Geology, National Museum of Wales, Cardiff CF1 3NP, United Kingdom.

Published 4 times a year by the Palaeontological Association.

Total circulation: 2,000.

Regular subscription: L70 individuals.

PALAIOS

ISSN 0883-1351

PALAIOS stresses the impact of life on Earth history and publishes papers on paleoecology, sedimentology, stratigraphy, paleoclimatology, biogeochemistry, paleoceanography, and their economic implications. Contains comprehensive articles, short papers, invited editorials, and essays devoted to the applications of paleontology in solving geologic problems.

Edited by David J. Bottjer.

Editorial address: David J. Bottjer, *Palaios*, Department of Geological Sciences, University of Southern California, Los Angeles, CA 90089.

Published 6 times a year by (SEPM) Society for Sedimentary Geology.

Total circulation: 1,300.

Regular subscription: $115.00.

Paleobiology

ISSN 0094-8373

Paleobiology publishes original contributions dealing with any aspect of biological paleontology, but with special emphasis on biological or paleobiological processes and patterns.

Edited by David L. Meyer and Arnold I. Miller.

Editorial address: David L. Meyer and Arnold I. Miller, *Paleobiology*, Department of Geology (ML 13), University of Cincinnati, Cincinnati, OH 45221-0013.

Published quarterly by the Paleontological Society.

Regular subscription: $35.00, $23.00 students, $65.00 institutions.

Photogrammetric Engineering & Remote Sensing

ISSN 0099-1112

PE&RS is devoted to the exchange of ideas and information about the applications of photogrammetry, remote sensing, and geographic information systems.

Edited by James B. Case.

Editorial address: American Society for Photogrammetry and Remote Sensing, 5410 Grosvenor Lane, Suite 210, Bethesda, MD 20814-2160.

Telephone: 301/493-0290.

Published monthly by the American Society for Photogrammetry and Remote Sensing.

Regular subscription: $120.00.

Physics and Chemistry of Minerals

ISSN 0342-1791

Physics and Chemistry of Minerals publishes fundamental physical and chemical studies on minerals and on solids related to minerals. Topics include crystal structures, solid-state spectroscopy, thermodynamics, kinetics and solid-state reactions, chemical bonding, and physical properties including fundamental parameters such as elasticity, compressibility, and thermal expansion. Such studies are important to geophysical and geochemical problems. Supports interdisciplinary work in mineralogy and physics or chemistry, with particular emphasis on applications of modern techniques and new theories.

Edited by I. Jackson, D. L. Kohlstedt, K. Langer, A. S. Marfunin.

Editorial address: I. Jackson, The Australian National University, Research School of Earth Sciences, Box 4, Canberra, Australia 2601.

Published 8 times a year by Springer International in cooperation with the International Mineralogical Association.

Regular subscription: $866.00 a year.

Physics of the Earth and Planetary Interiors

ISSN 0031-9201

Physics of the Earth and Planetary Interiors is devoted to the application of chemistry and physics to studies of the Earth's crust, mantle, and core and to the interiors of the planets.

Edited by D. Gubbins, D.E. Loper, J.-P. Poirier.

Editorial address: D.E. Loper, Florida State University, Geophysical Fluid Dynamics Institute, 18 Keen Building, Tallahassee, FL 32306.

Telephone: 904/644-6467.

Published by Elsevier Science Publishers.

Subscription: For rates, contact Elsevier Science B.V., Journal Department, P.O. Box 211, 1000 AE Amsterdam, The Netherlands.

Precambrian Research

ISSN 0301-9268

Precambrian Research publishes studies on all aspects of the early stages of the history and evolution of the Earth and its planetary neighbors.

Edited by B. Nagy and A. Kroener.

Editorial address: B. Nagy, Laboratory of Organic Geochemistry, Department of Geosciences, Gould-Simpson Building, The University of Arizona, Tucson, AZ 85721.

Fax: 602/621-2672 (Nagy).

Published 20 times a year by Elsevier Science Publishers.

Regular subscription: Dfl 1,805.

Quarterly Journal of Engineering Geology

ISSN 0481-2085

The *Journal* invites papers from all areas of the world on topics concerning geology as applied to civil engineering, mining practice, and water resources.

Edited by J.C. Cripps.

Editorial address: Staff Editor, Geological Society Publishing House, Unit 7, Brassmill Enterprise Centre, Brassmill Lane, Bath BA1 3JN, United Kingdom.

Published quarterly by Geological Society Publishing House.

Regular subscription: $245.00.

Quaternary Research

ISSN 0033-5894

Quaternary Research publishes interdisciplinary articles that deal with the Quaternary Period. Articles must have broad interest, high merit, and basic significance to more than one discipline (geology, paleontology, archeology, paleobotany, paleoecology, paleoclimatology, Quaternary oceanography, pedology, volcanology, and glaciology).

Edited by Stephen C. Porter.

Editorial address: Quaternary Research Center AK-60, University of Washington, Seattle, WA 98195.

Telephone: 206/543-1190.

Published 6 times a year by Academic Press.

Regular subscription: $218.00, $262.00 foreign.

Reviews of Geophysics

ISSN 8755-1209

The objectives of *Reviews of Geophysics* are to provide an overview of geophysics and the directions in which it is going and to serve as an integrating force in geophysics.

Edited by Alan Chave, Thomas Cravens, Kevin Furlong, Douglas Luther, Peter Molnar, James Smith, Richard Stolarski, Thomas Torgersen.

Editorial address: Alan Chave, Editor in Chief, Department of Geology and Geophysics, Woods Hole Oceanographic Institution, Woods Hole, MA 02543.

Published by the American Geophysical Union.

Regular subscription: $24.00 members, $250.00 institutions.

Science

ISSN 0036-8075

Science publishes research and review articles in all the sciences. News of recent international developments and research in all fields of science. Publishes original research results, reviews, and short features.

Edited by Daniel E. Koshland Jr.

Editorial address: 1333 H Street NW, Washington, D.C. 20005.

Telephone: 202/326-6500.

Published weekly by the American Association for the Advancement of Science.

Total circulation: 155,000.

Regular subscription: $87.00 a year, $205.00 institutions.

Sedimentary Geology

ISSN 0037-0738

Sedimentary Geology encompasses all aspects of research into sediments and sedimentary rocks, and at all spatial and temporal scales.

Edited by K.A.W. Crook, A.D. Miall, B.W. Sellwood.

Editorial address: Editorial office, *Sedimentary Geology*, P.O. Box 1930, 1000 BX Amsterdam, The Netherlands.

Published monthly by Elsevier Science Publishers.

Regular subscription: $1348.00 institutions.

Sedimentology

ISSN 0037-0746

Sedimentology publishes papers that deal with every aspect of sediments and sedimentary rocks.

Edited by J. E. Andrews, B. G. Jones, A. G. Plint.

Editorial address: J. E. Andrews, School of Environmental Sciences, University of East Anglia, Norwich NR4 7TJ, United Kingdom.

Telephone: 47/5521-3390.

Published 6 times a year by Blackwell Scientific Publications for the International Association of Sedimentologists.

Total circulation: 3,000.

Regular subscription: L185 a year, L205 foreign.

Soil Science

ISSN 0038-075X

Soil Science publishes original, authoritative, interdisciplinary research articles, as well as timely reviews, of interest to soil scientists, agronomists, environmental scientists, and soil-testing bureaus.

Edited by Robert L. Tate III.

Editorial address: Department of Environmental Sciences, Cook College, Rutgers University, P.O. Box 231, New Brunswick, NJ 08903.

Telephone: 908/932-9800.

Published monthly by Williams and Wilkins.

Total circulation: 6,000.

Regular subscription: $84.00 a year, individuals; $124.00 foreign.

Soil Science Society of America Journal

ISSN 0361-5995

The SSSA *Journal* publishes peer-reviewed, original research papers on soil physics, soil chemistry, soil biology and biochemistry, soil fertility and plant nutrition, pedology, soil and water management and conservation, forest and range soils, nutrient management and soil and plant analysis, soil mineralogy, and wetlands soils.

Edited by D. Keith Cassel.

Editorial address: Soil Science Society of America, 677 South Segoe Road, Madison, Wis., 53711.

Telephone: (608) 273-8080.

Published six times a year by SSSA.

Nonmember subscription: $108.00 a year, individuals; $120.00 foreign.

Tectonics

ISSN: 0278-7407

The editors of *Tectonics* welcome original scientific contributions in analytical, synthetic, and integrative tectonics. Papers are restricted to the structure and evolution of the terrestrial lithosphere with dominant emphasis on the continents.

Edited by Kevin Burke, American Geophysical Union, 2000 Florida Avenue, NW, Washington, DC 20009 and Sierd Cloetingh, Vrije Universiteit, Institute for Earth Studies, De Boelelaan 1085, 1081 HV Amsterdam, The Netherlands.

Published by the American Geophysical Union.

Regular subscription: $25.00 members, $250.00 institutions.

Tectonophysics

ISSN 0040-1951

Tectonophysics publishes original research studies and comprehensive reviews in the field of geotectonics and structural geology.

Edited by K. P. Furlong.

Editorial address: K. P. Furlong, Pennsylvania State University, Department of Geosciences, 439 Deike Building, University Park, PA 16802.

Telephone: 814/865-3620.

Published 56 times a year by Elsevier Science Publishers (14 vols.; 4 nos./vol.).

Regular subscription: Dfl 4,270 a year.

Water Resources Bulletin

ISSN 0043-1370

Publishes original papers covering water-resources issues. Includes litigation and legislation issues.

Edited by William Lord.

Editorial address: AWRA, 950 Herndon Parkway, Suite 300, Herndon, VA 22070-5528.

Telephone: 703/904-1225.

Circulation: 4,000.

Published 6 times a year by the American Water Resources Association.

Regular subscription: $110.00, $130 foreign.

Water Resources Research

ISSN 0043-1397

An interdisciplinary journal integrating research in the social and natural sciences of water. The editor of *WRR* invites original contributions in scientific hydrology; in the physical, chemical, and biological sciences; and in the social and policy sciences including economics, systems analysis, sociology, and law.

Edited by George Hornberger.

Editorial address: WRR, 2015 Ivy Road, Suite 407, Charlottesville, VA 22903.

Published by the American Geophysical Union.

Regular subscription: $115.00 members, $675.00 institutions.

AAPG Explorer
Box 979, Tulsa, OK 74101. 918/584-2555
(Vern Stefanic, Managing Editor).

AASP Newsletter
Robert L. Ravn, Sohio Petroleum Co., 50 Fremont St., San Francisco, CA 94105.
415/979-4981.

Acta Crystallographica
C. E. Bugg, BioCryst Pharmaceuticals Inc., 2190 Parkway Lake Drive,
Birmingham, AL 35244.

American Scientist
Rosalind Reid, P.O. Box 13975, Research Triangle Park, NC 27709.
919/549-0097.

Applied Earth Sciences
44 Portland Place, London, England W1N 4BR.
01/580-3802 (M. J. Jones, Editor).

Arctic and Alpine Research
Institute of Arctic and Alpine Research, Campus Box 450, University of Colorado,
Boulder, CO 80309.
303/492-3765 (Kathleen Salzberg).

Australian Journal of Earth Sciences
A. E. Cockbain, Geological Survey of Western Australia, 100 Plain Street, East
Perth, WAA 6005 Australia.
09/367-7037.

Bulletin of Canadian Petroleum Geology
Ashton Embry, Institute of Sedimentary and Petroleum Geology,
3303-33rd Street NW, Calgary, Alberta, Canada T2L 2A7 (I. Hutcheon, Editor).

Bulletin of Marine Science of the Gulf and Caribbean
Editor, Editorial Office, Bulletin of Marine Science, Rosenstiel School, University
of Miami, 4600 Rickenbacker Causeway, Miami, FL 33149.
305/361-4190 (William J. Richards, Editor).

Bulletin of the International Association of Engineering Geology

Laboratoire Central des Ponts et Chaussees, 58, Blvd. Lefebvre, 75732 Paris Cedex 15, France.

1/532-3179 (L. Primel, Editor).

Canadian Geotechnical Journal

J. I. Clark, Centre for Cold Ocean Resources Engineering, Memorial University of Newfoundland, St. John's, Newfoundland, Canada A1B 3X5.

The Compass

Don C. Steinker, Department of Geology, Bowling Green State University, Bowling Green, OH 43403.

419/372-7200.

Deep-Sea Research

John D. Milliman, Department of Geology and Geophysics, Woods Hole Oceanographic Institution, Woods Hole, MA 02543.

Earth, Moon, and Planets

V. Vanysek, Kluwer Academic Publishers, The Journals Editorial Office, Earth, Moon, and Planets, P.O. Box 17, 3300 AA Dordrecht, The Netherlands.

Earth-Science Reviews

Editorial Office, Earth-Science Reviews, Box 1930, 1000 BX Amsterdam, The Netherlands.

Earth Surface Processes

Prof. M. J. Kirby, School of Geography, University of Leeds, Leeds LS2 9JT, United Kingdom.

Estuarine, Coastal, and Shelf Science

Academic Press, Orlando, FL 32887.

305/345-4100.

Geographical Magazine

1 Kensington Gore, London, England SW7 2AR.

GeoJournal

Wolf Tietze, Magdeburger Strasse 17, D-38350 Helmstedt, Federal Republic of Germany.

0-53-51/72-33.

Geological Journal

A. E. Adams, Editor-in-Chief, Geological Journal, Department of Geology, University of Manchester M13 9PL, United Kingdom.

Geophysical Prospecting

Editorial Office, P.O. Box 298, 3700 AG Zeist, The Netherlands.

Geothermics

M. J. Lippmann, Lawrence Berkeley Laboratory, University of California, Berkeley, CA 94720.

Grana

Swedish Museum of Natural History, Palynological Laboratory, S-10405 Stockholm 50, Sweden.

08/158414 (Siwert Nilsson, Editor).

Icarus

Space Sciences Building, Cornell University, Ithaca, NY 14853.

607/255-4875 (Joseph A. Burns, Editor).

Isochron/West

John H. Schilling, Nevada Bureau of Mines, University of Nevada, Reno, NV 89557.

702/784-6691.

Journal of Applied Geophysics

Editorial Office, Journal of Applied Geophysics, Box 1930, 1000 BX Amsterdam, The Netherlands.

Journal of College Science Teaching

Lester G. Paldy, Director, Center for Science, Mathematics, and Technology Education, State University of New York at Stony Brook, Stony Brook, NY 11794-3733.

516/632-7075.

Journal of Geomagnetism and Geoelectricity

Prof. M. Kono, Terra Scientific Publishing, 302 Jiyugaoka Komatsu Bldg., 24-17 Midorigaoka 2-chome, Meguro-Ku, Tokyo 152, Japan.

Journal of Geophysics

C. Kisslinger, Cooperative Institute for Research in Environmental Sciences, University of Colorado, Boulder, CO 80302

Journal of Glaciology

International Glaciological Society, Lensfield Road, Cambridge, England CB2 1ER.
0223/355974.

Journal of Marine Research

Box 6666, Kline Geology Laboratory, Yale University, New Haven, CT 06511.
203/436-3715 (George Veronis, Editor).

Journal of Physics of the Earth

Center for Academic Publications, 2-4-16, Yayoi, Bunkyo-ku, Tokyo 113, Japan.
3817-5825.

Journal of the Geological Society of Japan

Faculty of Science, University of Tokyo, 7-3-1 Hongo, Bunkyo-ku, Tokyo 113, Japan.

Lithology and Mineral Resources

Plenum, 233 Spring Street, New York, NY 10013.
212/620-8468 (U. N. Kholodov, Editor).

Marine Geophysical Researches; an International Journal for the Study of the Earth Beneath the Sea

Bedford Institute of Oceanography, Box 1006, Dartmouth, Nova Scotia, Canada B2Y 4A2.
902/426-3549 (R. C. Searle, F. K. Duennebier, Editors).

Marine Geotechnology

Crane, Russak & Co., 3 East 44th Street, New York, NY 10017.
212/867-1490 (Ronald Chaney, Editor).

Mineralium Deposita

Institute fur Allgemeine und Angewandte Geologie der Universitat Munchen, Luisinstrasse 37, 8 Munchen 2, Federal Republic of Germany.
089/5203-247 (D. D. Klemm , H. J. Schneider, Editors).

The Mountain Geologist

Jim Schmoker, Executive Editor, 730 17th Street, Suite 350, Denver, CO 80202.
303/236-5794.

The NSS Bulletin: Journal of Caves and Karst Studies

National Speleological Society, Cave Avenue, Huntsville, AL 35810.
303/271-1073 (Andrew J. Flurkey, Editor).

New Zealand Journal of Geology and Geophysics

The Editor, New Zealand Journal of Geology and Geophysics, S & R Publishing,
 P.O. Box 399, Wellington, New Zealand.

(64/4-472-7421).

Oceanus

Oceanus Magazine, Woods Hole Oceanographic Institution, Woods Hole, MA
 02543.

The Palaeobotanist

Birbal Sahni Institute of Palaeobotany, 53 University Road, Lucknow 7, Lucknow,
 226007, India

(B.S. Venkatachala, Editor)

Palynology

Douglas J. Nichols, U. S. Geological Survey, Mail Stop 919, Box 25046, Denver,
 CO 80225.

303/236-5677 (David G. Goodman, Editor).

Physics and Chemistry of the Earth

Pergamon Press, Maxwell House, Fairview Park, Elmsford, NY 10523 (J. A.
 Pearce, Editor).

Physics of the Solid Earth

American Geophysical Union, 2000 Florida Avenue NW, Washington, D.C.
 20009.

202/462-6903.

Planetary and Space Science

C. P. McKay, NASA Space Science Division, Ames Research Center, Moffett Field,
 CA 94035.

415/604-6864.

Pure and Applied Geophysics

Renata Dmowska, Division of Applied Sciences, Harvard University, Pierce Hall, 29 Oxford Street, Cambridge, MA 02138.

617/495-3452.

Radio Science

American Geophysical Union, 2000 Florida Avenue NW, Washington, DC 20009.

Review of Palaeobotany and Palynology

Editorial Office, Review of Palaeobotany and Palynology, Box 1930, 1000 BX Amsterdam, The Netherlands.

Rock Mechanics and Rock Engineering

Springer-Verlag, 175 Fifth Avenue, New York, NY 10010 (K. Kovari, Editor).

Rocks and Minerals

Heldref Publications, 4000 Albemarle Street NW, Washington, D.C. 20016. 513/574-7142 (Marie Huizig, Editor).

Scottish Journal of Geology

The Editors, Scottish Journal of Geology, Geological Society Publishing House, Unit 7, Brassmill Enterprise Centre, Brassmill Lane, Bath, BAI 3JN, United Kingdom.

Space Science Reviews

C. de Jager, SRON Space Research Utrecht, Sorbonnelaan 2, 3584 CA Utrecht, The Netherlands.

Surveys in Geophysics

K.M. Creer, Department of Geophysics, University of Edinburgh, King's Buildings, Edinburgh, Scotland EH9 3JZ.

031/667-1081

Tellus

Tellus, Editorial Office, Arrhenius Laboratory, S-106 91 Stockholm, Sweden. 08/162000 (H. Lejenas, Editor).

World Oil

Robert E. Snyder, Gulf Publishing Co., P.O. Box 2608, Houston, TX 77252-2608.

Index